Digital Legacy

D1400776

Digital Legacy

Take Control
of Your
Online Afterlife

By
Daniel Sieberg
Rikard Steiber

SD

Published by Stonesong Digital, LLC
New York, NY USA

Digital Legacy: Take Control of Your Online Afterlife

Copyright © 2020 by Daniel Sieberg and Rikard Steiber

ISBN: 978-1-7362059-3-8 (KDP epub)
ISBN: 978-1-7362059-4-5 (KDP paperback)

10 9 8 7 6 S 4 3 2 1

First Edition, 2020

*For Natalie, Kylie, and Skye—my forever yesterday, today, and tomorrow.
And for my family—for their endless love, laughter, and inspiration.*
—Daniel Sieberg

*For Annika, Alexandra, Athena, and PaPi—the girls that rule my world
and already "hacked" all my accounts.*
—Rikard Steiber

Contents

Preface

Death.

They say you die twice—first, when you stop breathing, and then a second time, a bit later on, when somebody says your name for the last time.

There are few subjects more difficult to discuss or imagine than death. It's like we'd rather talk about anything else than the one universal experience we all share.

Yet, as we spend more and more of our time in the digital world working, relaxing, playing, and connecting online, there's an urgent need to overcome this trepidation and prepare for death from a digital-legacy standpoint.

Just as we arrange life insurance or trusts and wills, we need to plan for the postdigital life experience not only for ourselves but for the loved ones we leave behind—or else we run the risk of losing our digital legacy that includes our memories, our stories, and even our identities.

Worldwide, as of the writing of this book, there have been more than 1.3 million deaths due to the COVID-19 global pandemic, which brought degrees of loss and heartache into virtually all of our lives. The scale was beyond comprehension, whether you or your family lost loved ones or, like so many of us, watched in horror as the daily news presented its drumbeat of infections, hospitalizations, and death.

As those tragic events unfolded during months of lockdowns and quarantines, our digital worlds became a kind of sanctuary within our deserted islands. We relied on online shopping for everything from toilet paper to oat milk to running shoes, and we burrowed into social media to stay in touch with friends and family and help distract ourselves from the onslaught of negativity. We clung to our personal devices like life preservers, because maybe they were. The constant glow of our laptops and computer screens (and TVs) became the new campfire.

It was during this time, in the spring of 2020, that my father passed away, and I decided to start a new company called GoodTrust to help protect people's digital legacy. His death made me realize that we spend our lives creating memories, friendships, wealth, and assets that we want to pass on to our family. We want to be remembered. And because most things we want to pass on now are digital, there needs to be a plan on how to transfer our digital legacies. After doing more research, I discovered that it is extremely complicated to hand over digital assets, and very few websites had a process for managing this.

To learn more about this issue, I did a survey of one thousand people and asked, "Do you know what will happen to your digital stuff when you die?" A whopping 90 percent said that they had no idea, and 84 percent said they needed a solution for this. Their responses were not only the impetus for starting GoodTrust, but for the writing of this book.

I met Daniel Sieberg through a mutual friend and former colleague at Google. We both hailed from the tech space and believed in technology's power for change and good. Daniel was the author of the great read *The Digital Diet: The Four-Step Plan to Break Your Tech Addiction and Regain Balance in Your Life* (Crown, 2011) and had written numerous articles on technology. He loved the idea of this book and agreed to become the quarterback in making it a reality.

This book has several goals: to generate a critical level of awareness about the issues surrounding digital afterlife; to foster discussion about the challenges involved, from our often-uncomfortable relationships with tech companies to our inherent lack of trust; and to empower you to make informed decisions about what's possible and what's worthwhile with regard to digital afterlife and take positive steps forward.

The good news is that you have more control over your digital afterlife than you realize, and we hope that you not only learn from this book about where we are and how we got here, but that you're inspired to take action. There's nothing particularly enjoyable about a subject like death, but death comes for us all. And we need to be prepared, whether we like it or not.

As Albus Dumbledore of *Harry Potter* once said: "To the well-organized mind, death is but the next great adventure," and it is my hope that this book provides you with the foundation to begin that journey.

—Rikard Steiber, November 2020

CHAPTER 1

Choose the Present to Preserve Your Future

"We all die. The goal is not to live forever, the goal is to create something that will."

—Chuck Palhniuk, author of *Fight Club*

Imagine that every object in your life that matters to you, everything that is meaningful, is inside one room: family heirlooms, treasured photos, money, expensive items, inexpensive items, your choice. Pack it full. Don't leave anything behind.

Now imagine that same room suddenly has impenetrable walls and no doors, no windows, no way inside. And you're standing outside of it. Locked out.

Then you die.

Everything you cared about is trapped inside that room. No will or estate trust will ever note their inheritance. No one in your life now or in the future will ever possess them at your behest or understand their value or learn about their stories. There is no copy of these objects, and you were the only person who believed in the reason for their existence or appreciated their inherent value. And now they have essentially disappeared. It is as if they never existed!

How does that make you feel?

Hold that thought.

Now imagine the opposite.

Every bit of information there is of you in this world—everything from what you think, what you say, what you see, what you buy, and where you go—was available for *anyone* to browse after you've died. Some of your most private information. Available to strangers. Your children. Your business competitors. Theirs for the taking.

How does *that* make you feel? Would we like what others see? The mistakes. The imperfections. The poor choices. The *humanity*.

Both of these scenarios are a reality when it comes to our digital footprints: all the information that has been collected by computers while we're alive. Our digital life encompasses our every interaction with a screen, including social media comments made, Google photos taken, playlists streamed, blog posts written, articles read, food ordered, destinations driven, and money spent. This, coupled with other media available about us online—articles written about us, social media photos we're tagged in, and so on—forms our digital legacy.

Tech **TALK**

Digital footprint: The information that exists about us on the Internet as a result of our online activity.

Digital legacy: The electronic data that is left behind when we die.

Digital assets: Anything that exists in a digital format that comes with the right to use and may have either sentimental or monetary value. This includes documents, photos, money accounts, cryptocurrency, blogs, websites, applications, media, domains, photos, and so on.

WE'RE UNPREPARED

How much are we aware of our digital legacy? Minimally, but on the rise, according to recent data on the topic. In 2017, the Digital Legacy Association, a nonprofit group based in the United Kingdom, examined people's awareness of digital afterlife and their preparedness for it. The results of that study found that:

- A whopping 96.9 percent of respondents hadn't made a plan for their purchased digital assets once they die.

- 84.8 percent hadn't made any plans for any of their social media accounts following their death.
- 95.8 percent hadn't completed a "social media will" to document what they would like to happen to their social media profiles upon death.

For a society that spends so much of its life wrapped up in e-mails, online banking, social profiles, and the like, it is ironic that it has done almost nothing to address what happens to it all after death, although there is good news: the Digital Legacy Association also reports that between 2014 and 2017 the number of people making plans for their digital estate and their digital legacy is increasing.

We can often talk ourselves in or out of anything that feels new or different. And many of us are professional procrastinators on some level, especially when it comes to something that feels like it doesn't have a sense of real urgency. But there is an increasing imperative to address the issue of digital legacy not just in our own lives but for those we care about. If it helps you ease into taking action, then maybe think of helping to do it on someone else's behalf—and what will be lost if you don't jump into your own bits and bytes now.

TWELVE REASONS PEOPLE DON'T TAKE ACTION ON THEIR DIGITAL AFTERLIFE

1. **Death is depressing:** Virtually no one wants to think about it, let alone dwell on it. (Some also believe that the mere mention of death will jinx us and accelerate its arrival.) But while our physical body dies, our digital body goes on.
2. **Our digital clutter can seem overwhelming:** So many sites, so many logins and passwords, so many privacy policies. But there is a way through.
3. **It feels narcissistic:** We all die, so let's just leave behind whatever there is and not get too wrapped up in it. Yet, it is our responsibility to plan for what we leave behind, for our own legacies and for our loved ones.
4. **I'm not a celebrity:** Indeed, a public figure with millions of followers or fans faces additional pressure to keep parts of his or her life

private, and there may be a need to protect information from hackers looking to capitalize on something scurrilous. But hackers tend to be equal opportunity criminals. You may have fewer prying eyes, but your information is no less consequential to you. It is worth something to you, which means it's worth something to them.

5. **I don't give away that much information about myself online:** You'd be surprised. An e-mail address here and an annual income there adds up. And behind the scenes they could be swapping information and stitching together a profile of you in the abstract.

6. **Nothing I do online is worth saving:** Are your digital photos, e-mails, and documents any less meaningful than your photo albums, letters, and paperwork of the physical world? And what about your domain or any cryptocurrency? The rise in cryptocurrency like Bitcoin or Ethereum can mean hundreds or thousands, if not millions of dollars in value. (For more on cryptocurrency, see Chapter 7.) Even massively multiplayer online role-playing game (MMORPG) characters and digital items can sell for thousands of dollars.

7. **No one will care:** Try looking at your online life through someone else's eyes. Ask your partner or your children or your friends what they might want to retain or preserve? Also, they may be left footing the bill with ongoing subscription services or lose out with any digital assets that should be passed on.

8. **Tech companies don't want to help:** Although tech companies don't yet make it as easy or robust as they should, there is a growing sense of needing to better address this global challenge. And virtually every online site or service or app today has some policy in place to help with the passing of a loved one. It just takes a little time to learn more.

9. **Once I'm gone, it won't matter if people dig up my online past:** It could matter far more than you think. Think about all those "private" Facebook messages you wrote in the heat of the moment. Your entire online life can be made available to anyone if you don't take action with clear directives. Devoid of context or understanding, there could be lingering concerns or confusion about whatever content is uncovered or arguments about who has access to what.

10. **Anything I want to delete will never really be gone anyway:** Yes and no. If you want to eliminate your online life and destroy

any traces you leave behind, then you can take action to request that outcome by contacting the companies involved with the site or requesting that it be removed upon your death. On the other hand, doing nothing will almost certainly guarantee a dispute by those left behind over whether your online life should be deleted or preserved or shared. It is best to make sure that your digital legacy is not only handled to your liking, but that it is complete and respectful.

11. **There are so many legal hurdles and potential costs that it doesn't seem worth it:** Although sites, services, and apps don't seem to make it easy for an individual to embark on this journey, it's further complicated by indecision on who can help with legal issues. The costs of managing a digital legacy are not insignificant depending on what you need to obtain (e.g., a court order and power of attorney), and the options are often vague or intractably filled with legalese, but there's a way through the confusion and complication.

12. **I have time to think about this:** Death can be sudden and unexpected, as many learned during the recent COVID-19 pandemic. And tragedy is exacerbated when there are no instructions for others to follow after a person is gone.

Quick**QUIZ**

Ask yourself:

- Do you have social media accounts of any kind that capture much of your life in photos, videos, or text?
- Do you wonder what happens to your financial accounts or important documents saved on a computer or remote server after you pass away?
- Have you ever worried that friends or family members might discover your search history?
- Do you worry that the words you've written in e-mails or texts when you're angry, sad, or rejected (or inebriated) might fall into the wrong hands?
- Have you ever wondered what your children will think—or your children's children will think—when they come across photos of you or information about you online when you're gone?

- Do you spend most of your day online creating anything that reflects something about who you are?

If you've answered yes to any of these questions, then you care about your digital legacy.

SMART THINKING

If you need further convincing that you care about your digital life, try this exercise:

- **Step 1:** Give your smartphone to an acquaintance to hold. How does that feel? Stressful? Fearful? You may instantly wonder if your password will hold, if the facial recognition lock really works, or if they can sense what's really in your smartphone just by holding it.
- **Step 2:** Do the same thing, except, this time, unlock your smartphone first. How do you feel now? More stressed? Praying for no screen notifications at that moment, which would offer sudden insight into our digital world?

People love their smartphones. According to 2019 data from Pew Research, 81 percent of Americans own a smartphone. That was up from 35 percent in 2011. By comparison, it took about fifty years for that same level of penetration for the television.

More than 71 percent of us either sleep with our smartphones on the nightstand, have it in bed with us, or even hold it while sleeping, according to Sleep.org. We check them before turning on the television and rely on them for everything from shopping to communication to socializing to news to entertainment. According to research by the tech care company Asurion, Americans check their phones upward of ninety-six times per day (and that's on the low end for some people) or an average of once every 9 to 10 minutes. (If you're wondering where you fall in the obsessive-compulsive smartphone owner scale, you can try an experiment using an app like QualityTime or BreakFree.) Our smartphones have become our constant companions and sometimes mean more to us than our real-life companions. A 2020 study by Reviews.org found that 45 percent of respondents would rather give up sex than their smartphone!

But did you ever think about why you're so attached? Yes, there are biological forces at play, involving chemicals and hormones (dopamine, cortisol) and those exciting little beeps or rings or constant obligations that let us know someone wants to interact with us, but our smartphones also serve as our mobile portals. They represent not only our connection to others, but our connection to ourselves, our digital lives, technological connective tissue that doesn't disappear once we die.

Think About It

- If someone looked through all the information you have in your smartphone, would they "know you"?

SOCIAL DILEMMAS

There are more than thirty million dead people with profiles on Facebook. Chances are, one day you will be one of them—it is estimated that by the year 2100, there will be roughly 3.6 billion dead profiles. Soon dead Facebook users will outnumber the living. Every day, more than ten thousand deceased Facebook users could be:

- Friend requested
- Tagged in a photo
- Wished a happy birthday

You probably know someone who has passed away, and yet his or her profile and feed remains, with friends posting remembrances for the deceased's special occasions. We see dead people everywhere: Facebook, Pinterest, Instagram, Twitter, Snapchat, YouTube, LinkedIn, and so on. Our physical bodies eventually die, but our digital bodies, formed by our online behaviors while we're alive, live on in perpetuity, if we allow them to.

A rather dystopian view of the impact of social media in our lives was front and center in a 2020 Netflix documentary titled *The Social Dilemma*. The film, spearheaded by the Center for Humane Technology, explores this lack of awareness many people have for their digital lives and how and why their personal information is used, often against them—how the technology that we've all come to know and love and depend on also manipulates and monetizes us. (Critics laud its important message, but some have cited a lack of deeper solutions.)

But our digital lives are not just limited to social media. They're wider than we think. How many of these sites contain information or digital assets that are meaningful to you and/or could be valuable to anyone you leave behind?

- Amazon
- Apple
- Coinbase
- E-Trade
- eBay
- Etsy
- Google
- Hulu
- Match
- Microsoft
- Netflix
- Paypal
- Photobucket
- Reddit
- Shutterfly
- Slack
- Spotify
- TikTok
- Tinder
- Venmo
- Vimeo
- Yahoo

And there are many, many more. Think about the dozens of restaurant and retail apps and/or sites you interact with. Have you thought about what information you willingly share? And that information, unless you stipulate otherwise, will remain available long after you're gone.

Did You Know . . .?

When social media sites such as Facebook and Instagram present you with "On This Day" or "Memory" posts from years ago, they are tapping into the human desire to connect with the past.

FINANCIAL RISKS

The issue of digital legacy is not just about who, what, and where, but also about how much. Ownership of digital assets can represent sizable value, and there may be financial risk and loss if there is no clear line of provenance. For example, in 2013, a survey by Internet security firm McAfee found that the average person in Canada valued their digital assets—entertainment files, video game purchases, e-books, movies, TV shows, and so on—at more than $32,000.

Even if you believe your digital life is valued more or less than that, the point is that it has value, and that value can be diminished, like a stock that plummets. For example, think of all the recurring subscriptions you currently have—Netflix, Amazon Prime, Weight Watchers, gym memberships, home security systems, cloud storage, and so on. A recent survey found that, on average, Americans spend about $237 monthly on these types of subscription services. If your loved ones aren't aware of those subscriptions or are not easily able to turn them off, there will be a considerable financial loss to your estate when you die.

———————

There's a lot to consider with our digital afterlife: the quantity of our information, the quality, its financial worth, its emotional worth.

Think of it as a correlated relationship: the more you know about digital afterlife, the safer you can make the time you spend online today, and the more control you'll have on how you'll be remembered tomorrow. Remember, what you leave behind in the digital world reflects your human story, and there's no way to undo your digital legacy after you're gone.

Did You Know . . . ?

There are:

- 3.5+ billion Google searches per day (equal to more than two trillion searches per year).

- 1.79 billion people active on Facebook every day (and five new profiles created every second).
- 500 million daily active Instagram users.
- More than a million Tinder swipes every minute.
- Thirteen new songs added to Spotify in every new 60-second interval.
- 600 new page edits to Wikipedia every minute.
- 500,000 tweets every minute.

THREE KEY TAKEAWAYS

- Your digital legacy is more than you might think: it can include correspondence and communications (historical value); digital artifacts, memories, and mementos (sentimental value); and digital assets and subscriptions (monetary value).
- Our real lives and our digital lives have merged into one experience. It's time we accept this new world order.
- You *can* take action and shouldn't let the process limit your desire to preserve what's important. After all, your digital afterlife is unique to you.

CHAPTER 2

~~~~~~~~

How the World Became Digital
There Is No Going Back

"Any sufficiently advanced technology is indistinguishable from magic."

—Arthur C. Clarke

Throughout their history, human beings have sought to preserve their existence and experiences. From cave drawings and petroglyphs to stone tablets, oral retellings, and written books, they have told stories and offered guideposts to help them and others understand their place in the world (and also themselves), share knowledge and wisdom, and pass on perspectives and lessons to others. Dating as far back as the Neolithic era of 5000 BCE, humans also literally weaved stories into the artistic design of things like baskets and rugs to pass them on to future generations. It is that concept of weaving stories that eventually led to technological innovations we can see today.

ORIGIN STORY: FULL OF HOLES

The genesis of programmable computing goes back to at least the late eighteenth and early nineteenth century with weaver and merchant Joseph Marie Jacquard in France. Jacquard managed to automate steam-powered weaving looms and create hole-punched cards with differing patterns. Those cards could then generate textiles based on the corresponding pattern (similar inspiration led to player pianos that fed the notes into the instrument). A few decades later, English mathematician and computer pioneer Charles Babbage's rudimentary but effective tabulation machine (the difference engine, which was never fully completed) applied steam power for gears and crankshafts.

The work of both Jacquard and Babbage would serve as lightbulb moments for census clerk Herman Hollerith of New York. Hollerith combined these ideas with electricity to speed up the tedious counting process, and his invention was put to work during the 1890 U.S. Census (and reportedly saved the government $5 million).

Later, these hole-punch-reading developments led to Konrad Zuse's invention in Berlin in 1936. A civil engineer and scientist, Zuse created a programmable digital device called the Z1, which used metal pins, plates, and old film, that could add and subtract. His early models were reportedly destroyed during World War II, but it could be considered a computational embryo of what we know today. (And maybe the reason there were "hanging chads" in 2000.)

THE 1940s: UNLOCKING THE CODE

The first substantial computer was the Electrical Numerical Integrator and Calculator (ENIAC), which dates back to World World II and is credited to John Mauchly; its primary function was to aid with ballistics analysis at scale. The machines back then could hardly power today's icon or a spreadsheet, but the seeds were planted for how humans would relate to these machines and the role they would play in people's lives by helping them overcome mental limitations by doing things like processing complex calculations or sorting huge volumes of data.

It didn't take long for businesses to seize on the advantage these machines provided in efficiency and productivity to manage payrolls, reports, and inventories. Eventually, scientists even realized the potential to both create and crack encryption, which was instrumental during World War II with machines like the Enigma and Colossus.

The bottom line: an immutable link between man and machine had begun.

THE 1970s: EXPONENTIAL GROWTH

By the 1970s, the arc of computing had gone from small in physical size and limited in actual capability to large in physical size and substantial in terms of actual capability. But with the invention of the microprocessor in 1972—following the invention of the integrated circuit chip of the late 1950s—the concept of a computer started to shift from a machine that had grown to the size of a room or a city block to something more portable and still powerful. It then became possible to imagine these machines sold and used in offices in a more personal capacity, including the Xerox Alto, which could print documents and even send e-mails. What's more, these new machines didn't require coders or scientists to understand how to interact with them, which opened the door to mass consumption and distribution.

The 1970s also represented a surge in computers becoming part of popular culture and the zeitgeist thanks to video games like Pong in 1972, coveted pocket calculators from Hewlett-Packard, the popularity of magazines devoted to DIY tinkerers like *Popular Electronics*, and a burst of Hollywood entertainment with computers and technology playing a central role, such as *Star Wars* (1977), *Alien* (1979), and *Westworld* (1973).

The bottom line: we were hooked.

Tech TALK: Early Edition

IT: Information technology (IT) is a term often attributed to Babbage's creation as the dawn of outsourcing human expertise and advancement to machines. IT now applies to all manner of how data is transmitted, stored, received, and processed as well as to a broader reference of distribution devices like computers, televisions, and cell phone systems. It's also now the go-to department for many people who need help with their work computers.

Vacuum tubes: These eye-catching but fragile encasements are glass tubes with the gas removed to create a vacuum. That allows electrons to flow and are often used as a switch or amplifier. They started appearing in all manner of electronics from radios to TVs to telephone systems before they were replaced by the more durable transistor. Some techno and audiophiles still prefer the way their devices function based on this early design.

Transistors: The origin of the transistor can be explained in some ways through the creation of the word (credited to engineer John R. Pierce). It combines the notion of *transfer* or *transconductance* with *varistor*, which can further be broken down to *varying resistance*. Transistors are essentially semiconductors that allow electronic signals or power to be amplified or switched more efficiently and at a much smaller scale than vacuum tubes.

Moore's law: Credited to American engineer Gordon Moore in 1965, Moore's law is a prediction based on empirical production projections that the number of transistors able to fit on a dense integrated microchip would double roughly every two years. (In other words, the technology would advance so rapidly as to shrink them down per nanometer of surface area.) Since that time, his prediction has largely held true though slowed somewhat as the scale has gotten so microscopic in nature. This forecast ultimately aided in the planning and development of devices we know today like smartphones, which now contain millions of transistors, resistors, and capacitors instead of hundreds.

THE 1980s: THIS TIME IT'S PERSONAL

By the 1980s, the rise of personal computers began, thanks to companies like Apple, Radio Shack, MITS, Commodore, and eventually IBM. Personal computers, or PCs as we know them, started to have a practical purpose in people's lives and allowed for new ways to create and to communicate and discover. We could print reports, articles, and even books; we could create spreadsheets and organize data better, and we could manipulate information and images into something personal or more valuable.

As individuals, we started to gain control of the technology and apply its capabilities in unique and personalized ways. Integrated circuits and microprocessors that represented thousands of bits in your hand were replaced by silicon computer chips that stored millions of bits in something the size of a fingernail. And herein lies one of the greatest correlations to the convenience of preserving stories, legacies, and memories: data storage (see page 21). All that digital stuff had to go somewhere, even if it was mostly out of sight, out of mind.

Tech TALK: Building the Brain

Memory: The earliest examples of computers, while impressive in their designs and innovation, lacked a foundational part of computing as we know it today: how to harness processing of information (electricity) in real time at higher speeds and what to do with the creation of any data. You can think of it in terms of the human brain: there's memory we need for immediate tasks (short-term to-do lists) and memory we rely on for later (stories, life experiences). For computers, it became essential to develop memory for logic and calculations as the applications for them evolved at scale.

Processing: At a basic level, the processing of data or information (or words) involves the collection and manipulation of items with a change that's detectable to an observer. The same can be said about the processing of instructions fed into a computer, which requires energy and power to do so in an efficient manner. The better the computer's central processing unit (CPU), which is measured in million instructions per second (MIPS), or supercomputers at floating-point operations per second (FLOPS), the better it can tackle large-scale challenges like forecasting climate change or quantum computing or molecular modeling.

Bits and bytes: The use of bits and bytes was once more commonplace with the size of computer files, but these days we use terms like "meg" or "gig" or even "terra," which are all multiples for the unit of a byte. There are 8 bits in every byte, and thus 8 megabits (Mb) in every megabyte (MB). Internet speeds are still usually measured in megabits per second or Mbps and not MBps since the data transfer happens one bit at a time. For example, a relatively small file like 38 MB would take a 38-Mbps connection 8 seconds to complete the transfer. Or about as long as it took to read the last two sentences.

Yottabytes: A yottabyte (YB) is a measure of theoretical storage capacity equal to 2 to the 80th power bytes, or approximately 1,000 zettabytes, a trillion terabytes (TB), or a million trillion megabytes. Approximately 1,024 yottabytes make up a brontobyte. The prefix "yotta" is based on the Greek letter "iota." In sum, that's alotta data. or In sum, that's a yotta data.

THE 1990s: A WIRED WORLD

We filled up these storage options so much that, by the 1990s, the synthetic and symbiotic relationship humans had with machines had accelerated to a whole new level. And we needed even more data storage in the form of external hard drives. We still had a binary relationship to computers: we logged off or we turned them off, and lived most of our lives in the physical, or analog, world. Then . . .

Enter the Internet in a big way, thanks to companies like AOL, CompuServe, Prodigy, and MSN (Microsoft).

Once the Internet—and, by extension, the World Wide Web, created by Sir Tim Berners-Lee at CERN in 1989 and made more widely available by 1993—spread to homes and offices around the world, there were new reasons to be online and new ways to store, create, and preserve: scanners, faxes, digital cameras, handycams, MP3 players, and e-book readers. And with more tools came a greater attachment to technology. We started to remain online for longer periods of time, exchanging e-mails or looking stuff up, because we could. We felt a sense of connection to people outside our neighborhoods and even outside our towns and countries. Technology had expanded our lives but had not yet taken them over.

Did You Know . . . ?

Every person generated roughly 1.7 MB of data per second in 2020. As a comparison, we have individually gone from saving thimbles of information in the 1980s to swimming pools by the 2020s.

THE 2000s: BORN DIGITAL

In the 2000s came machine learning—and big data. We couldn't stop generating 1s and 0s. We got Wi-Fi in our homes and carried portable Wi-Fi devices in our pockets. The information explosion was released forever. Big data was everywhere.

Even the word "big" in big data doesn't do it justice during the past twenty years. It was so big that we needed to program computers just to handle where to put all the data. And it was at this point, in 2000, that researchers like Peter Lyman and Hal Varian (the latter went on to be the chief economist at Google) declared that data was becoming democratized and dominant—so much so, that most and eventually all textual information (and images) would be "born digital."

Beyond the need to store and process and manage all this data was the need to *understand* it. How to find the needle in a stack of needles? Thus, machine learning became a critical component of big data to mine through it all and uncover the most salient, valuable, and actionable nuggets of information for any company or business model. Of course, this created social issues related to consumers as merely data points at scale for making money and countless ethical challenges on how to preserve that data, protect it, combat bias, and decide who can or should share it. Today, there is an entire regulatory framework around the privacy of big data (see Chapter 3).

INTRODUCTION OF E-MAIL

The first e-mail ever recorded dates back to 1971 and is credited to Ray Tomlinson, who was working as a computer engineer with BBN Technologies, which was a company hired to help build the U.S. government's Advanced Research Projects Agency Network (ARPANET) and the forerunner of the Internet. Tomlinson says he created the e-mail (electronic mail) software, dubbed SNDMSG, as a side project to see if computers could exchange messages between people and not just send electronic communication to a numbered mailbox, which had been done in the past. The first recipient of his e-mail revelation? Tomlinson himself. And he recalled in interviews that it was likely something pretty benign like *QWERTIOP* as a simple test to see if it worked.

Oh, how e-mail has grown since then.

According to a McKinsey study in 2019, the average worker spends 28 percent of their day buried in e-mail. As a full-time employee, that represents about 2.6 hours and 120 messages received every day. And all those e-mails represent a digital asset. In 2005, Google introduced Gmail with its 1 GB of storage, enabling users to save correspondence, with the added benefit of searching it. In many ways, e-mail has come to truly replace "snail" mail as not only the means and destination where we exchange the most meaningful correspondence but our e-mail address is now akin to our address in the virtual world—unique to us and a prized, protected, and unique identifier. Many of us are also guilty of excitedly watching our in-box for a certain message to arrive instead of walking to the mailbox on the curb. (In both cases, we still deal with junk mail or spam.)

INTRODUCTION OF THE SMARTPHONE/TABLET

The mobile nature of computing had its origins with devices like the Palm Pilot, which also helped coin a new acronym: PDA or personal digital assistant. In 1992, Jeff Hawkins founded Palm Computing and reportedly would carry a small block of wood and broken chopstick into investor meetings to help them understand his vision for something portable and interactive with a stylus.

Prior to the Palm Pilot, Hawkins had invented a slightly larger device called the GridPad in 1989, which is often referred to as the earliest mainstream tablet computer (fictional ones had appeared in popular culture, such as writer Arthur C. Clarke's NewsPad featured in the 1968 movie *2001: A Space Odyssey* and even the Calculation Pad described by Isaac Asimov in his 1951 novel *Foundation*).

The Apple Newton tablet was released in 1993 and featured its own operating system but never seemed to quite catch on. Microsoft launched the Tablet PC in 2000 with great fanfare, but that also failed to gain traction. In both cases, the devices were limited by issues like their weight, battery consumption, and limited applications. It was only later with Apple's iPad in 2010 that the tablet became more popular and now leads the product category, alongside ones from the likes of Samsung, Microsoft, and Amazon.

INTRODUCTION OF THE SEARCH ENGINE

Google, which launched in 1998 and now amounts to more than 86 percent of all online searches worldwide, is clearly entrusted with mind-boggling amounts of our data. It was Google's page-ranking algorithm that checked how many pages linked back to a particular site that proved to be the most effective and popular, but it certainly wasn't the first to allow us to search for information online with precursors like Infoseek, Lycos, and Altavista. Today other top search engines include the likes of Bing (Microsoft), Baidu (primarily in China), Yahoo!, and Yandex (primarily in Russia). All of them served or continue to serve as our personal encyclopedias, but there may be few of us who really want all that data preserved forever or reflected back to us or passed on to the next generation after we die. At least not without our permission.

INTRODUCTION OF SOCIAL MEDIA

Six Degrees, MySpace, Friendster, Blogs, Facebook, Twitter, Instagram, Pinterest, TikTok, QZone, VKontakte—the list goes on and on.

By 2017, some 2.5 billion of us were on social media. By 2020, that number had increased by nearly 50 percent to 3.8 billion people—that's more than half the entire world's population. And with social media, we're no longer just socializing or peering over the virtual fences of our neighbors: we buy and sell stuff, watch videos, and discover new activities or hobbies.

Social media has become our daily journal, a record of our lives. On average, we spend about 2 hours and 24 minutes per day on its various sites. That's an increase of an hour a day, or up 62.5 percent, since 2012, while e-mail has largely plateaued or even decreased.

Over the course of our lives, that collective time spent on social media adds up to six years and eight months and growing. For some people that might be nearly one-tenth of their entire life and, statistically, is only surpassed by watching TV and sleeping. Of those social media experiences, the top destinations are Facebook, YouTube, Snapchat, Instagram, and Twitter. Others popular in the United States and elsewhere include WhatsApp, LinkedIn, and Pinterest.

Many of those users are eighteen to thirty-four years old or Gen Z, in today's vernacular, and an increasing percentage of Gen Z are also using online games like Fortnite with 350 million players, Roblox with 150 million players, and Minecraft with 126 million players. Games have become the new place for digital social interaction and connection with new friends.

Did You Know . . .?

During the average lifetime, we'll spend five years and four months on social media. That is:

- More time than we'll spend eating, drinking, grooming, and socializing (in real life).
- Enough time to fly to the moon and back thirty-two times.
- Enough time to walk the Great Wall of China 3.5 times.
- Enough time to watch the entire *The Simpsons* series 215 times.
- Enough time to run more than 10,000 marathons or climb Mt. Everest thirty-two times.
- For most of us, approximately one-fourteenth of our entire time alive.

INTRODUCTION OF APPS

The concept of a singular program running on a floppy disk or hard drive or elsewhere on a computing machine stimulated an entire industry built around software for countless applications. Applications! Indeed, the now-ubiquitous term "app" grew from *applications*, which started appearing on mobile devices as a portable version of the software or programs on computers, laptops, and other machines. They *applied* to more specific aspects of our lives as we used our technology on the go (e.g., the weather app or a news app or a flight app). Companies that created software for computers had discovered fertile ground to tailor these apps for mobile users and convert existing ones into a new format. The mobile app market was worth $106 billion in 2018 and is expected to reach upward of $407 billion by 2026. A new gold rush was born.

THE HISTORY OF DATA STORAGE

As we expect more and more from computers and task them with saving and recalling more information, we need effective and efficient ways to preserve all that data. Our human brains are limited in myriad ways, even though we process a billion billion calculations per second, and especially when it comes to saving and recalling huge amounts of information (except for maybe *Jeopardy!* champion Ken Jennings).

Data storage represents the methods and technologies that capture and retain digital information. Storage is a key component of digital devices, and as the volume of data fed into the machines increased, so too did the need to put it somewhere. But unlike storing anything physical, this type of digital storage needed to conceal a colossal amount of digital stuff—documents, files, lists, data, and so on—now in our lives. Eventually, those personal computers became capable of storing much more, and processing power allowed for faster speeds to do so.

To solve for this exploding mountain of data, scientists and engineers eventually developed three kinds of digital storage: electromagnetic, optical, and silicon based. Remembering back to the early punch cards of the nineteenth century to process information, the net result was still a piece of paper, which then had to be stored and organized somewhere. That option became highly impractical over time.

By 1932, Austrian developer and pioneer Gustav Tauschek had invented the magnetic drum (about 16 inches long), which spun at 12,500 revolutions per minute and could store information on its read-write heads on the axis. These electric pulses were then converted into a series of binary digits. Although electromagnetic storage was incredibly useful for code breakers during World War II and remained relevant during the 1950s and 1960s, it eventually became used only as an auxiliary backup. And its storage capabilities wouldn't exactly help with the digital packrats we've all become: the first drums held about 48 KB or about five formatted .doc files.

Magnetic storage evolved over time to become spinning magnetic platters, which are referred to as hard-disk drives (HDD). First introduced by IBM in 1956, the original HDD was the size of a refrigerator and weighed more than a ton, though it could handle 3.75 MB of data—or roughly 45 seconds of your favorite low-resolution cat video. Eventually, HDD was replaced by solid-state drives (SDD), which are still used today, and models like one from Samsung at 2.5 inches can hold 16 TB, which is equivalent to about four million digital photos taken with a 12-MB camera. But, of course, that still wouldn't be enough as the pace of evolution charges forward.

The rise and fall of smaller magnetic storage for consumers like floppy disks, cassette tapes, and VHS (R.I.P.) led to optical storage like CDs and DVDs (also mostly R.I.P.), and introduced the world to the power of lasers. Lasers! Developed in 1982 by Sony and Phillips, the CD could hold more data than a personal computer's hard drive by shining a laser at the surface of the disc. The storage capacity of CDs was roughly 650–700 MB, which amounts to basically the number of digital songs in a music album. By 1995, DVDs expanded that capability to 1.46 GB of storage space or enough to hold an average movie. We later enjoyed Blu-ray discs starting in 2003 (named after the relatively short wavelength of the blue laser to read the higher-capacity discs), though today digital files and streaming media through cloud-based storage have largely replaced all that.

This leads us to our current merger of silicon-based storage devices and the connectivity of the Internet, coupled with the cloud-based nature of it, which allows digital information to be available anywhere on any device. Today, our digital footprints are tracked and captured through remote and behemoth databases that are housed in storage centers that are an average of 100,000 square feet and as big as 6.3 million square feet (about 110 football fields), the largest one in the world in Langfang, China, from Range International Information Group, or 3.5 million square feet like the one from Switch SuperNAP outside Las Vegas, Nevada, which is the largest in the United States. Each one is able to instantaneously sort and process information and transmit it to our portable devices and other machines that are accessible to users through on-demand cloud services and Internet connectivity all the way to users at the "edge" of the networks, no matter their location.

Are you seeing a trend here? The technological advancements of storage and memory have allowed humans to virtually extend their limited mental capabilities and create a bottomless pit of digital stuff. Why become a selective photographer when you can just take twenty photos in a few seconds and choose the best one? Why carry around a few books when you can store dozens of them on one Amazon Kindle? Why bother cleaning out your e-mail in-box when you can organize them into digital files and archive them?

Without the inconvenience of clutter, human beings have become digital archivists. All of our information is neatly tucked away, out of sight, in folders, hard drives, and the cloud. As the processing power and personal devices grew smaller, the ability to store them needed to grow larger.

Did You Know . . .?

In 2017, IBM released a report that found that 90 percent of all data in the world had been created in the past two years. And if that weren't mind-blowing enough, the amount of data that humans created annually far exceeds the creation of all human records dating up to the early 2000s. That's a staggering amount of information.

WHERE WE ARE NOW: AT THE BEGINNING
OF A NEW REALITY

Today, our lives have become so dependent on the Internet and connectivity that we can barely imagine life without it. 5G, the fifth-generation technology standard for broadband cellular networks, will be more widespread soon, and with it will come a whole new era of experiences that will revolutionize the way we live, work, and play.

5G's ultralow latency (the delay in the transfer of data) and huge bandwidth makes it the ideal technology for edge computing and data transmission, leveraging supercomputing power and unlimited cloud capability to deliver our favorite content in high resolution in real time.

The main apps for 5G may be video and virtual and augmented reality (see box). With a trillion sensors from 100 billion connected devices gathering data across autonomous cars, phones, drones, wearables, cameras, and so on, all connected to a low-latency high-speed network, you'll be able to know anything you want, anytime, anywhere, and query that data for answers and insights.

ARTIFICIAL INTELLIGENCE

In 1985, twenty-two-year-old world chess champion Garry Kasparov went to Hamburg, Germany, to play thirty-two simultaneous games against the strongest chess computers of the day. How many matches did he win? All thirty-two. And no one was surprised. At the time, who could imagine that a chess computer would beat a human grand master?

Twelve years later, in 1997, Kasparov, still a world champion, lost to IBM's Big Blue, and the world was amazed. (Today, most of the free mobile chess apps would beat Big Blue of old.)

Say hi to AI, or its full name: artificial intelligence. To explain AI, it helps to think of a spectrum: on one end is a machine-learning component whereby humans (and sometimes computers or programs on their own) need to sort, process, and evaluate data to glean insights or patterns. In the middle of the spectrum, that output then becomes part of an algorithm or even a series of algorithms called a neural network or networks. And at the other end of the spectrum is the AI or how closely it resembles or exceeds the mind of a human when it interacts with people.

Recently AlphaZero, a descendant of the AI program from a company called DeepMind (originally acquired by Google in 2014) that first conquered the board game Go, taught itself to play a number of other games at a superhuman level. After 8 hours of self-play, the program bested the AI that first beat the human world Go champion; after 4 hours of training, it beat the current world champion chess-playing program, Stockfish. Today, AI can teach itself things without any human programming, and the speed of technology advancement is about to increase significantly.

Although these days AI is primarily used for specific applications such as playing chess, driving cars, or buying stocks, the recent advances in AI and machine learning will be a core component in bringing forth the next industrial revolution and new experiences. Although we are still some time away from general self-aware AI systems like the ones in Hollywood movies such as 1968's *2001: A Space Odyssey*, soon AI will be woven into our everyday life. The evidence is already there: when we ask Google, Siri, and Alexa for help, when we let Tesla drive us home. It is very likely that you soon will be able to have your personal AI assistant that will be created from all of your data online. This AI will act on your behalf when you are too busy answering e-mails, chat messages, booking meetings, and more. This AI could live forever online. (For more on the future of technology as it relates to digital afterlife, see Chapter 10.)

Tech TALK

- **Augmented reality (AR):** An interactive experience that adds digital elements to a live view often by using the camera on a smartphone, Magic Leap, or Microsoft HoloLens. Examples of AR experiences include Snapchat spectacles and the game Pokémon Go.

- **Virtual reality (VR):** A complete immersion experience that shuts out the physical world. Using VR devices such as HTC Vive or Oculus Rift, users can be transported into a number of real-world and imagined environments such as the middle of a squawking penguin colony or even the back of a dragon.
- **Extended reality (XR):** An umbrella term that covers all of the various technologies that enhance our senses, whether they're providing additional information about the actual world or creating totally simulated worlds for us to experience. It includes AR and VR.

BRAVER NEW WORLD

According to We Are Social's 2019 Digital Report, there are approximately 4.4 billion Internet users worldwide generating 2.5 quintillion bytes of data each day and spending an average of 6 hours and 42 minutes online. That means, collectively, in 2019–2020 human beings spent 1.2 billion years online (perhaps even more if you factor in the COVID-19 pandemic). Additionally:

- There are five billion searches every day online.
- More than 500 million tweets are posted.
- Some 294 billion e-mails are sent.
- There are more than 4.7 trillion photos digitally stored somewhere in the world.
- If you wanted to download all the data from the Internet, it would take you more than 181 million years.

As technology has evolved and machines have "learned" and increased their ability to interpret mass amounts of data, this has empowered companies to sort, analyze, and apply this data in many different ways, including, most recently, spotting patterns across millions of data sets to spot COVID-19 symptoms in CT scans and help trace the location of people who may be infected.

Additionally, companies now build entire business models around collecting this data, providing content or search results or advertising that corresponds to it. There are better and better predictive algorithms, enhanced understanding of human behavior, and innovative opportunities in smart cities and agriculture and transportation and so much more.

Technology is feeding those primitive and hardwired needs for storytelling and community and knowledge in various ways. But does all this excitement come with a price? With a drive toward efficiency, effectiveness, and enhancement, ethics is often left behind, which means the price may be on our digital heads.

THREE KEY TAKEAWAYS

- Humans are complicated and diverse creatures, but we all seem to share a number of traits: we create stuff, we need to store stuff, and we need to remember stuff. And technology has paradoxically become both the solution and the problem.
- The pace of technological innovation is only accelerating—in some cases, at an exponential rate—and, by many accounts, we are heading squarely toward a greater merger of human and machine.
- We are our digital lives. Whether it's social interactions, the digital bits and bytes we keep and share, or the ways in which we interact with all of it, the value in both a monetary and sentimental sense has reached previously unimagined levels.

CHAPTER 3

Eye of the Beholder
Entrusting Your Digital Legacy to Others

"None of us knows what might happen the next minute, yet still we go forward. Because we trust. Because we have Faith."

—Paulo Coelho, author of *The Alchemist*

In today's tech-driven world, we routinely trust myriad companies, staffed with employees we've never met, with everything from our memories to our money. We sign up online or through an app, and we barely blink when asked for our personal data, our preferences, and our logins and passwords. We place our faith in them because they offer us what we need in the moment—convenience, or access to information or services they provide. We may accept that our data is somehow used in ways we don't love to think about, but push that all aside in exchange for whatever the site or service offers. Or we rarely question their intentions or their efforts behind the scenes unless something awful or public takes place.

THINGS WE GIVE AWAY

1. Our thoughts: through search engines such as Google or Bing
2. Our friends: through social media like Facebook or LinkedIn
3. Our banking details/credit card numbers: through online retailers such as Amazon or GrubHub

4. Our locations: via smartphones
5. Our destinations: via Uber or Google Maps
6. Our DNA: through companies such as 23andMe and Ancestry

PRIVACY IS AN ILLUSION

Although we're routinely presented with privacy policies and legalese on the definitions and limitations of our digital relationships, we rarely look too closely. According to a 2017 Deloitte survey, only 9 percent of us ever read privacy policies and terms of service—and only 3 percent of eighteen- to thirty-four-year-olds do. The reasons vary from "it takes too much time" (the average privacy policy is 2,500 words) to "I really don't care what's involved" to "I don't understand what it all means anyway."

A Pew Research study from 2014 found that roughly half of Americans don't even know what a privacy policy is, believing that they're put in place to ensure the confidentiality of their information (false) versus protecting the liability of the company itself (true).

Privacy experts like Mark Rasch, the former chief privacy evangelist for Verizon Communications and vice president, deputy general counsel, and chief privacy and data security office at SAIC, will tell you that privacy is largely an illusion no matter where you live and that privacy laws, especially in the United States, are a patchwork quilt of liability protection and a confusing state of where onus is placed. Ultimately, our online interactions come down to a simple choice: we either do or we don't. We simply click *Accept*. And if we don't, then there's no access. So we click and hope for the best, craving real-time experiences, but ignoring the implications today, tomorrow, and, especially, after we die.

Tech TALK

Crowdsourcing efforts including sites such as ToS (Terms of Service), DR (Didn't Read), and TLDRLegal (Too Long Didn't Read Legal) have sprung up to make privacy policies more digestible since the vast majority of us make Faustian bargains without blinking an eye. These sites basically use the same principles as something like Wikipedia to enlist and moderate the interpretation of complicated privacy policies and try to explain them in terms that resonate with a busy but interested consumer.

Tech **TIP**

Read the fine print. Even if it doesn't make sense. The privacy and data policies listed by any tech company are a blueprint to what you should expect. Although there are questions as to whether they are transparent enough, provide enough feedback and guidance to your questions, and respect as a customer, determining what you expect is the first step.

THEY WILL BE WATCHING YOU

If you have any business presence at all online, then you know the value of trying to understand your audience. From start-ups to Fortune 50 companies, many businesses use some combination of analytics tools from Google Analytics to Nielsen to everything in between to figure out who their customers are. The data is represented through dashboards or insights gleaned from tiny pieces of code dropped on your computer, called "cookies," that reveal activities, such as how quickly you come and go on a site (bounce rate), where you linger (average time on page), your shopping preferences, and how many individuals (unique visitors) arrive each day and keep coming back.

What do companies want to know about us? With Google, for example, it's a closer look at what you're (really) thinking when you conduct a search (check out Google Trends to see what people all over the world are thinking in an anonymized fashion). For Facebook and other social media, it's about learning your interests and your beliefs and your friends (and their overlapping information). Your Internet service provider knows which sites you like to surf (unless you cloak yourself in a virtual private network (VPN), but even then it's not always 100 percent secure). Your credit-card company and other sites know your shopping transactions and what you like to buy. And it goes on from there.

Here is a list compiled by *Inc.* magazine in 2019 of the top twenty-two things tech companies (and really any company with an online presence) want to know about us at an even more granular level:

1. Personal information like name, gender identity, birthday, contact info
2. Location and address, along with where you spend much of your time
3. Relationship status, including whether you just started a family
4. Work status and income level (to focus the ads you see)
5. Educational background
6. Race/ethnicity
7. Religious and political beliefs
8. Facial-recognition data (often used for surveillance, but don't forget that option for unlocking your smartphone!)
9. Financial and banking information
10. IP (internet protocol) address—your unique identifier online
11. Communications and chat history
12. Calendar and how you spend your time
13. Search history
14. Entertainment choices and media consumed (including movies, books, music, etc.)
15. Web-browsing history
16. Social media behavior
17. Purchase history
18. Fitness and health data
19. Clicked ads
20. Posts on social media that you opt to hide (there are no secrets online)
21. Devices used to connect
22. Voice data, such as for Amazon Alexa or Google Home

This is, of course, valuable data for any business. It's not so much about "spying" on people but about adjusting a retail strategy or branding, or testing new features to learn about the response. Our behaviors as consumers are what drives companies to change the design of their site to keep us "in the store" longer, to buy more stuff, to contact any sales department, or to sign up for a service. There is now a whole data-driven discipline associated with interpreting these insights and reflecting those back to marketers and others within the company to inform various decisions.

The reality is that we leave behind a maze of digital footprints as we virtually saunter through our online journey. There's essentially nowhere we go online without someone or some company somewhere knowing and retaining some (limited but revealing) information about us—the more aggregated data over time, the more valuable it becomes. That identifiable information usually comprises fairly benign metrics such as geographic location down to a metro area, and it's less about exactly who you are and more about how you behave. And, yet, as it relates to digital afterlife, who we are and how we behave online become somewhat synonymous. We are the summation of where and how—and how often—we clicked.

Think About It

In the physical world, we accept that a business owner will observe our activity while we are inside his or her establishment—a store, a restaurant—and we act accordingly. Doesn't it stand to reason that we will be observed on some level when we "visit" those businesses' websites? What's more, brick-and-mortar business owners don't know any more than they can "see" when we enter the door, but they might know more than what they can "see" when we visit their online site.

TRUST IS EARNED

Trust is a funny thing. We don't give it easily. And sometimes we give it for reasons that have to do with how we *feel* rather than what we *know* to be true. Take, for example, the following Top 10 list from Morning Consult, a global data intelligence company, which listed the most trusted firms and products in the United States in 2020:

1. United States Postal Service (USPS)
2. Amazon
3. Google
4. PayPal
5. The Weather Channel
6. Chick-Fil-A
7. Hershey
8. UPS
9. Cheerios (General Mills)
10. M&Ms (Mars candy)

It's probably not surprising that the U.S. Postal Service, an American institution, tops this list. Since Benjamin Franklin was the first postmaster general in 1775, the U.S. Postal Service has earned our trust, because our letters and packages arrive when we expect them to. More or less. (Sign of the times: In August 2020, Facebook signed a lease for 730,000 square feet of the former U.S. Postal Service building in New York City and will inhabit all of the office space for its headquarters.) Also, the four food companies/products listed have relatively long histories with consumers, and the three tech companies—Amazon, Google, and PayPal—have been around for at least twenty-plus years and have become integral parts of our lives. We feel we *know* them.

And yet the Morning Consult survey also showed that a new generation of consumers are not quite so ready to adopt the brand love of their parents. Younger consumers are generally more skeptical of corporate America, hold higher ethical standards for brands, and are more distrustful of brands across the board. So while brand trust is relatively strong today, there is an additional challenge facing brands in the future.

Takeaway: Trust is earned over time, difficult to build.

Additionally, brand relationships vary by level or tier: How you trust a brand with your money is different than how you trust one with your shopping delivery, weather forecast, quality of your milk chocolate, or, in the case of digital afterlife, your trusted information. Therefore, our trust is granted on certain terms, depending on the relationship and our expectations. When companies fail us, or disappoint us, that trust can erode.

Takeaway: Trust is conditional and easy to lose.

BAD REPS

David Polgar is the founder and executive director of All Tech Is Human, a nonprofit organization that seeks to inform and inspire the next generation of responsible technology leaders. Polgar has interviewed founders, venture capitalists, executives, and visionaries in the tech space and has found that when any enterprise fails or develops a bad reputation, it may be because:

- Senior executives broke their trust either internally with their colleagues or externally with their customers or both. They stopped delivering on their promises.
- Senior executives may have behaved immorally.
- Data may have been compromised.
- A combination of all three.

When trust is breached, there is rebuilding necessary to restore faith in the product or service.

IN TECH WE TRUST?

According to a 2015 survey from the Public Affairs Council, there is a pattern within different industries when it comes to trust. The following industries are listed from most trusted to least trusted:

1. Manufacturing
2. Technology
3. Large retailers
4. Food and beverage

5. Automobile
6. Energy
7. Financial institutions
8. Health insurance
9. Pharmaceutical

The tech industry rated high on this list, second only to manufacturing. Perhaps it's because approximately two-thirds of those surveyed said they would trust companies more if they treated employees well, and tech companies have long offered perks and benefits to employees, from free food to extra time off. Also, 54 percent of respondents said they would like to see a net benefit to society from a company's efforts, and tech firms often support community initiatives: Twitter recently pledged $1 million to the Committee to Protect Journalists and the International Women's Media Foundation, and the ride-sharing service Via discounted private rides for people who needed to travel during the pandemic by 20 percent.

But it's not all good news for tech. The Trust Barometer Report from the public relations firm Edelman in early 2020 found technology, manufacturing, and retail at the top of the list, but the technology sector slipping during the past few years. Respondents cited a sense that regulation is coming to major tech companies but also confusion about whether the government officials involved actually understand the issues. Misinformation and accusations on various platforms have also complicated the trust relationship between consumer and tech companies. According to a PwC survey of consumers, 76 percent of people described sharing their personal information with tech companies as a "necessary evil" in today's modern economy. Additionally, 52 percent of customers reported they would leave one company in favor of another if it means they would protect their data better.

Additionally, although a 2015 Public Affairs Council survey found that some 71 percent of Americans believed that tech companies had a positive impact on the country, by late 2019 the Pew Research

Center reported that number had plunged to 50 percent, and the negative view of tech firms rose from 17 percent to 33 percent. (A 2019 Dice Insights survey broke it down even further by looking specifically at Facebook. According to its findings, 86 percent of respondents did not trust Facebook.)

Data breaches may be part of the reason for that growing mistrust. According to a 2020 survey from cybersecurity company Varonis, consumers are more likely to mistrust a newer tech company like Uber after a data breach versus one like Target; perhaps companies with a longer track record have an easier time restoring that trust while newer ones need to act to prevent losing out to a crowded field of competitors. In both cases, Varonis reported that 85 percent of respondents said they would spread their opinion to others, meaning the damage can ripple well beyond those people directly affected.

There have been many data breaches in the past decade, and during that time the number of consumers affected has reached into the tens of millions. Some of the most high-profile incidents tracked and ranked according to financial impact and loss by CSO magazine include:

- Adobe, October 2013 (153 million user records)
- Adult Friend Finder, October 2016 (412.2 million user records)
- Canva, May 2019 (137 million user accounts)
- Dubsmash, December 2018 (162 million user accounts)
- eBay, May 2014 (145 million users)
- Equifax, July 2017 (147.9 million consumers)
- Heartland Payment Systems, March 2008 (134 million credit cards exposed)
- LinkedIn, 2012 and 2016 (165 million user accounts)
- Marriott International, 2014–2018 (500 million customers)
- My Fitness Pal, February 2018 (150 million user accounts)

There are any number of reasons why a security breach takes place: from human error to weak points along the cyber security supply chain to outdated protocols within the system. And the motivations range from financial gain to political hacking to general mischief.

Tech TIPS: When Deciding Whether or not to Trust a Company

1. Do a thorough review of the company's online presence (site, social media, etc.) and look for any glaring spelling or grammatical errors. We've all made typos, but a company's public-facing presence should be professional and precise. And with social media, perhaps look closer at how (or even if) the company interacts with any comments.

2. Look for contact information and customer-service support, such as a phone number or e-mail, and reach out to the company. You might take the opportunity to ask a few questions or comment on its business efforts, but in either case gauge how those interactions feel to you. Is the company answering your questions or comments directly and taking you seriously?

3. Is there a privacy policy? Examine it as closely as possible, and compare it to other sites and companies you already use and trust. You might also look it up on one of the crowdsourcing sites to get a more digestible version.

4. Search for any reputable news stories about the company and learn more about the founders and executives who run it (sometimes more information can be found about them on LinkedIn).

5. On the more technical side, you might consider looking up the company's domain registration through a website such as WhoIs .net, although that doesn't always provide much detail and isn't always accurate. And ensure the company has site encryption measures in place, such as *https* (or a lock symbol) in the URL.

LEAP OF FAITH

There's really no magic formula to create trust. We take a leap of faith every day when we interact with technology. Most of the time that works, and we don't feel the need to worry, but sometimes it doesn't. Certainly, a data breach doesn't foster trust in online security, but security professionals suggest that we need to treat them as a fact of life.

Also, the trust doesn't end when we stop using an app or a program or a service. That trail of digital DNA, which when pieced back together, is the sum of our digital existence, whether we like it or not, and we must not only start thinking about all those tiny pieces that we're leaving behind but how to take steps to preserve and protect that digital legacy for ourselves and others. Most companies are not designed or optimized for handling your digital afterlife. After all, once you're dead you're not really much of a customer anymore, are you?

THREE KEY TAKEAWAYS

- Companies the world over use our data in order to do business. They watch our online behaviors and use those behaviors to inform various business decisions.
- Consumers put their trust in companies to safely and securely live their digital lives. But while a company may "feel" trustworthy, there is still an onus on us to conduct due diligence. Trust, but verify.
- Data is only as safe as its weakest link and often that's unknowable by consumers. *Caveat emptor*, or let the buyer beware, with our digital lives, and take the time to occasionally take stock of where all your data lives. Know thy (digital) self.

CHAPTER 4

'Till Death Do Us Part
Evaluate and Designate

"As much as it's great to plug in and be connected and feel limitless, there is no real total opposite of that in our society anymore. There is no way to totally shut it off or opt out."

—Andie Diemer, journalist

We may have succeeded at connecting the world, but what are we leaving behind? Although our online time may seem random and perhaps not all that useful to us, there are compelling reasons to save these actions and records if we choose to do so.

> ## Tech **TIP**
>
> Sorting through your digital assets will take time and a lot of thought. Consider these questions when assessing their value:
>
> - What if one of those sites suddenly went out of business?
> - How would you feel if it were gone?
> - Could it be easily recovered?
> - Does some of this information mean a lot to people in your friends and family network?
> - How would others around you feel if it were deleted?
> - Do you have a clear sense of what happens to any of that content whether it's through their mismanagement or your inaction?
> - If you passed away, could someone find your content or data?

Tech **TIP**

When compiling any digital asset list, it might help to score or rank each item to help illustrate their importance in your life. You can do it by category: since we often have more than one social media account, rank each of them in terms of value and content. You might put Facebook at or near the top, since it contains photos and contact information for friends, which may be necessary for reaching out after someone passes on. That could be followed by Instagram, if you treat it as a kind of personal story, and then maybe YouTube if you've uploaded a lot of videos there to be saved or shared later. Ask yourself: If I lost access to any of my digital life while I'm still alive, which loss would be the most devastating? That will help you decide what's most important to you.

WHAT TO KEEP, WHAT TO TOSS

The place to begin may be with a little list making and a little soul searching. Jot down as much of your digital life that you can think of and how it's accessed, including:

- E-mail.
- Documents and files.
- Photos stored in the cloud (and which ones are most valuable among what are potentially thousands!).
- Social media profiles (Instagram, Facebook, TikTok, Snap).
- Financial, trading, and insurance accounts.
- Registered domain names.
- Digital music playlists (note: the streaming song files belong to the site/artist).
- Professional accomplishments (e.g., LinkedIn).
- Movie and TV sites such as Apple+, Disney+, Hulu (note: the streaming files belong to the site, but sharing or saving someone's favorites might be meaningful).
- Amazon or other shopping accounts.
- Medium content.
- Dating sites (Match, Tinder, Bumble, etc.).

- YouTube videos.
- News sites.

Once you've figured out what types of digital assets you'd like to keep, you may want to consult with friends or family members. What should you leave behind? Ask members of your family— parents, siblings, children, cousins, anyone who is important to you—what they'd like to have of your digital assets when we're gone. If they do not understand what is available to them, pose these questions:

- What if . . . instead of being forced to rummage through boxes of my grainy black-and-white photos and you could relive my experiences in a VR setting?
- What if . . . instead of trying to trace your 23AndMe DNA to third and fourth cousins or a piecemeal approach to Ancestry. com, you could easily search through a private collection of our family history in a visual and comprehensive way?
- What if . . . I could share my favorite playlists and movies with your children and grandchildren so they could learn more about my creative side and entertainment preferences?

Preserving and extrapolating these insights could help everyone learn so much more about each other and ourselves, and pass that wisdom and learned experience on to our children and others.

Tech **TALK:** Postmortem Definitions

Account holder: The user of an online service or profile holder while they're alive; refers to the decedent if the user passes away and leaves an account open.

Digital directive: An instruction or decision that is made by an account holder that indicates how they want their information to be managed after they pass away.

Steward: The next-of-kin, survivor, fiduciary, or beneficiary of the account holder; the person who has been designated to manage the account after the account holder has passed away.

ENTRUSTING OUR LEGACIES

The global death care market is expected to top $102.4 billion in 2020, according to a Business Research Company report. That's a compound annual growth rate of only .1 percent, which the report states is largely due to the economic slowdown of the pandemic and measures to contain it, but it forecasts a 7 percent compound annual growth rate in 2021 to reach $124.8 billion in 2023.

Between life insurance, estate planning, and funeral homes, there is more than $1 trillion spent annually in the United States alone. And with an aging demographic online—for example, 72 percent of Facebook users are between fifty and sixty-four—there is a growing need to address both the physical and digital legacy of their lives.

To hack through the forest of data, how do we get started? It's like those giant standalone maps at an airport or shopping mall: if you don't know where "You Are Here" is, then the rest of the information isn't particularly helpful.

Unfortunately, as we discussed in Chapter 3, the starting point isn't just one place: our digital lives are spread across many different companies. And that means interacting with twenty or more companies that have twenty or more (and different) processes with twenty or more privacy policies.

How do you get started? Here are two ways to handle digital afterlife:

1. Do it yourself.
   a. **Pro:** You're in control.
   b. **Con:** It may take more time and perhaps more money as you navigate all the complexities.
2. Hire a company to do it for you.
   a. **Pro:** You have peace of mind as you allow a third party to conduct the process and go through all the legal hurdles.
   b. **Con:** You need to ensure they have your best interests at heart.

The way companies handle digital afterlife care with consumers was the subject of three fellows from the Aspen Institute Tech Policy

Hub in 2020: Liv Erickson, Donnelly Krum, and Matthew Schroeder. Their report, The Digital Afterlife Project, provides a series of recommendations and guidelines for any entity to better manage this issue and adopt three primary principles along the way: safe, simple, respectful. The basic categories for their research into effective account management involve deactivation, memorialization, and (data) stewardship:

1. **Deactivation features:** The most widely applicable to close down the account, delete its content, or transfer its data and hide the profile from the public.
2. **Memorialization features:** To indicate that the account holder has passed away.
3. **Stewardship features:** To allow a survivor to act in place of the account holder for certain tasks, generally to curate a mourning community or carry on the deceased user's work.

The Digital Afterlife Project authors also recommended a few critical components to the broader tech industry and any company doing business online:

- Ensure that the features are simple to understand, and don't provide any unnecessary steps, such as asking people to fax or mail something instead of making it possible to upload documents. Minimize the requirement for legal documents.
- Allow users to have as much choice as possible—without overwhelming them—in terms of options for what to do with an account or data.
- Do everything possible to share the account and data through legal channels without placing that burden on the shoulders of the survivors (e.g., whenever possible it's best not to transfer logins and passwords to someone else).
- Deploy ethical and compassionate language and instructions that make it clear you understand their confusion or concerns.
- Maintain regular and thorough communication with the survivors to ease their anxiety and uncertainties.

There are also companies that combine digital legacy care with a more holistic focus that includes:

- Sharing a final message or recordings with loved ones.
- Password managers and vaults and legal services to help parse through all the unwieldy stuff.
- Regular updates provided with trusted friends and family members or "deputies" (see box) along the way like a status bar for how everything is proceeding.
- Traditional wills and medical directives.

A WORD ABOUT DEPUTIES

When assigning "deputies," with their duties, such as knowing the logins and passwords to your online accounts or tasking them with instructions on how to find them when the time comes, only select trusted individuals. (Think of the word "deputy" in terms of the Wild West era; it's someone who is always right at your side and always has your back.) Choosing a deputy is like choosing the executor of your estate or a best man or maid of honor at a wedding. These are people who will not only keep your information safe, but also will potentially act on your behalf to sort through your digital content after you pass on. They are individuals with your best interests at heart and the right amount of discretion when needed.

If there are still too many bumps in the road, then a company might also offer "white glove" and VIP support to really provide a one-to-one experience and assuage any lingering concerns around what's needed and what's possible.

Think of it as a one-stop shop for all your afterlife needs. In that way, companies in this space can hope to tap into another of the tenets of trust with transparency and hold themselves accountable to accelerate the process as much as possible.

Any company in this space of managing digital afterlife is preserving the most precious commodity of all—our memories. There is likely no second chance if that information is misused, mismanaged, or misappropriated, which means it must be treated with the utmost respect and care. It's the equivalent of handing someone a cherished photo album or irreplaceable baby pictures and asking them to hold onto it for an indefinite period of time. The time is now to ensure that our data is saved in a manner that's respectful, ethical, and relevant.

QUESTIONS TO CONSIDER WHEN CHOOSING A DIGITAL AFTERLIFE COMPANY

- **Accessibility:** Do they provide a holistic list of sites to help manage?
- **Accountability:** Do they offer a comprehensive privacy and security policy? Are they endorsed somewhere or run by individuals with a public standing?
- **Availability:** Do they make it possible to easily contact them?
- **Cost:** The price may vary depending on the efforts undertaken, but ideally it would be less than working with a lawyer and perhaps in the neighborhood of $100 to $250.
- **Listening:** Do they provide an empathic and respectful approach to your concerns and questions?
- **Longevity:** Do they appear to be a fly-by-night operation without serious investment in this issue?
- **Personalization:** Does it feel like their consideration of your situation is unique to you?
- **Testimonials/reviews:** Can you tell if others have used them before? Do they have a social media presence?
- **Transparency:** Are they candid about the challenges and hurdles involved? Do they provide regular updates?

THREE KEY TAKEAWAYS

- Take stock of your digital assets. Make a list of what's important to you and what is important to others to preserve or delete.
- There are essentially two roads to take when embarking on a digital afterlife journey: you can go solo, or enlist the help of a company in the digital afterlife space. Either way, there is still an

onus on you to do your research and understand what's possible, who to trust, and what makes the most sense for your situation or those of others.

- Have a list of attributes for any digital afterlife firm with which you're considering a relationship. Just as you would entrust the care of your children with only a vetted individual or company, so should you entrust your memories.

CHAPTER 5

Hard Decisions
Organize and Prioritize

"For every minute spent organizing, an hour is earned."

—Benjamin Franklin, American philosopher,
statesman, inventor

Whether you are preparing for your own digital afterlife or that of another, whether you are taking on this journey yourself or partnering with a company, the process begins the same way: with organization and moving forward. According to a survey in 2015 by password-manager service Dashlane, the average person has roughly ninety online accounts that cover everything from e-mail to streaming media to banking to cloud storage to news, which means it's time to make some hard decisions about who will care for your digital legacy and to ensure they have what they need.

WHAT YOU NEED TO DO (AT A HIGH LEVEL)

1. Provide access.
2. Supply (find) authentication documents.
3. Consider next steps: funerals and memorial services.

1. Access

Access to your digital life not only includes a list of all your digital interactions (as discussed in Chapter 4), but making sure that list makes its way to the right people and the right places:

- **Assign legacy contacts:** This is different from a power of attorney or anything involved with legal considerations. It means entrusting someone with at least knowing what digital assets matter to you and how to find them.
- **Provide passwords:** Handing over your passwords can be uncomfortable and antithetical to everything we've heard about never sharing them. But it is a necessary step in taking control of your digital afterlife. Think of it as providing the key to your imaginary room, no different from giving a key to your physical home to someone in case of an emergency. Just be judicious.

> **Tech TIP**
>
> To lessen the anxiety of sharing your passwords, even to a trusted person, consider doing a password swap: have that person entrust you with their passwords in return. This will be helpful not only when one of you dies, but also during those moments in your life when you can't remember your passwords.

> **Tech TIP**
>
> Folders can be your best friend when it comes to digital organization. Create folders of important photos or e-mails, or sort them by name so that upon your death your loved ones will know which folders were meant for them.

2. Authentication Documents

There will always be a need to compile certain authentication documents before you can access personal data stored somewhere. Although any company will want to ensure that gathering them is as low-lift as possible, that doesn't mean it's without some doing on your part. For example, the following items will most likely be needed to access a variety of the sites, depending on the origin:

- **Court order:** Depending on the type of information (e.g., Google Photos), it can be required for access and is unique to each site. One court order does not work for all sites.
- **Customer full name:** Legal name as provided to the site.
- **Legacy contact:** A trusted point of contact—often the next of kin, but not always.
- **Power of attorney:** To make decisions on the person's behalf.
- **User birth certificate:** May be required if the site needs additional ID.
- **User death certificate:** The executor of the will and estate will have access.
- **User e-mail:** Best to use your primary one, though you may also be able to provide a backup.
- **User ID:** A driver's license or passport to verify authenticity.
- **User obituary:** Not always necessary but worth having (for tips on obituaries, see page 51).
- **User payment information:** Credit card information could be stored and may serve as identification material.
- **User phone number:** Usually a cell number.
- **User as executor:** Empowering an executor may be necessary.

Tech **TIP**

Because death certificates are essential for logging into a deceased person's profile and routinely required by tech companies in order to gain access to an account, instruct loved ones to keep your death certificate in a safe place upon your passing.

3. Consider Next Steps: Funerals and Memorial Services

Once upon a time, the idea of gathering for a last good-bye, whether for a funeral, a wake, or an interment, meant physically huddling together in one place. That was in large part because we lived and died locally. Due to advancements in communication and transportation, the branches of family trees are now stretched beyond neighborhoods and cities, and spread across states and countries.

The concept of funerals and memorial services has also expanded beyond walls and borders and is another aspect of preparation for our digital legacy. An entire cottage industry has sprung up in recent years devoted to this effort with companies such as Remembering. Live, Gather, GatheringUs, or GoodTrust (our company), all of which offer various services to create online memorial services.

For example, Remembering.Live offers a facilitator for a video conference, and you can upload shareable videos and photos and even have the ability to play a person's favorite songs. Gather streams the in-person service online to those people who can't attend in person, while GatheringUs offers a hybrid of online and offline services to accommodate people who join in person or through an Internet connection, while also translating the service into Spanish or French, if desired. All of these companies also offer a concierge option; prices for these services average around $1,400, depending on the options you choose. GoodTrust also makes it possible to pass along a favorite playlist and record a last good-bye to share with loved ones.

These kinds of online tributes allow people to participate in the memorial service in a meaningful way while remaining in the comfort of their home. When Broadway actor Nick Cordero died of complications from COVID-19 in July 2020, his wife, Amanda Kloots, put together an extensive online tribute the following September that included commentary from family and friends, music, and plenty of stories about his life. Proceeds from the event went to the Save the Music Foundation. (For more on memorialization, see Chapter 7.)

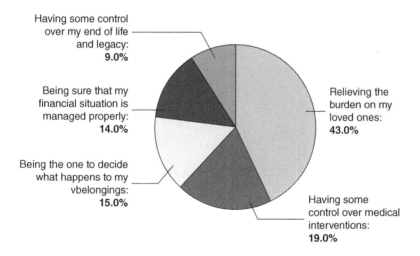

Tech TALK: Artifacts for an Online Service

- **Videos of good-bye messages:** From those in attendance, those who couldn't join, or those created by you for your own funeral.
- **Digital maps:** A recreation of all the places you visited around the world, perhaps highlighted with photos of various occasions.
- **Favorite songs:** A playlist of digital music that can be played throughout the service and even shared with people to access afterward.
- **Obituary:** A written document or video that can be shared with the participants during the service.

GETTING OUR AFFAIRS IN ORDER

According to a 2019 survey of adults, ages fifty-five and older, by Merrill Lynch Wealth Management Services, these are the primary benefits expressed when it came to getting their final affairs in order:

Having some control over my end of life and legacy: **9.0%**

Being sure that my financial situation is managed properly: **14.0%**

Being the one to decide what happens to my vbelongings: **15.0%**

Relieving the burden on my loved ones: **43.0%**

Having some control over medical interventions: **19.0%**

Once we have everything in place—we've organized our digital assets, appointed our deputies, and arranged our final send-off—it's time to move onto the next step: protection. We need to protect our digital legacy from inaccessibility and inadvertent deletion, as well as from those who would do it harm.

THREE KEY TAKEAWAYS

- Decide who you trust with your most personal digital stuff, and have a conversation with them about your wishes and intentions.
- Lean into the daunting process of organizing documents, and do your best to keep a record of everything relevant to preserving someone's legacy (whether offline or online).
- Consider what makes the most sense for a celebration of your life, or the lives of others, and explore the possibility that a virtual option might be the most pragmatic and inclusive.

CHAPTER 6

Don't Become a Victim
Outsmarting the Hackers

"Don't always trust what you see on social media. Even salt looks like sugar."

—Anonymous

Fraud. Scammers. Deception.

According to 2013 data from fraud prevention firm ID Analytics, every year the identities of roughly 2.5 million deceased Americans are accessed by thieves looking to do everything from fraudulently create credit card accounts and apply for loans to open cell phone accounts and acquire other services. Even more staggering, the firm reports that 800,000 deceased people within that 2.5 million are deliberately targeted. That's about 2,200 per day!

Sadly, even with the right laws in place, the malicious and predatory elements of the Internet will always be there. And as more people leave a trace of their lives online after death, there are more ways for the malevolent among us to prey on that vulnerability.

How? It happens in increments. A cyber security incident may take place on one particular site, and the attackers combine that data (e.g., name, address, and birth date) with the information they can learn from a social media page, which could inform them about passwords (we often choose phrases from our life, such as pet names or addresses), and this provides the bad guys with a fuller picture of how to either impersonate individuals or gain unauthorized access to their accounts.

Did You Know . . .?

According to cyber security research firm PurpleSec, close to 98 percent of security breaches are the result of human interactions (and thus human frailty) called "social engineering," or basically trying to trick or fool people into providing information through phony e-mails or phone calls. If someone at a business or bank isn't aware that an individual has died, then they may be vulnerable to sharing information in this way.

Tech **TIP:** Password Dos and Don'ts

Using a password manager can alleviate the need to remember so many different passwords while adding a layer of convenience. However, whether you use one or not, you will still need a strong password for any particular service, and there are general rules to follow:

1. **Don't** use something obvious, such as your birthday or your spouse's birthday, or your children's names. And please, please don't use "password" as your password!
2. **Do** change your passwords often—maybe as often as once a month or every few months, certainly whenever you receive a notification from a site to do so and if there's been a security breach.
3. **Do** use two-factor authentication, which means a strong password plus confirming your identity with a text message or e-mail. This is sometimes required by a site or app; other times, you'll need to opt in. It's an extra step but worth it.
4. **Do** incorporate uppercase letters, numbers, and symbols to make it more difficult for someone to guess or for a "bot" to randomly hack your account.
5. **Do** consider using a passphrase instead of a password, for example, horsemountainDenver1995!#, something that is unique to your life and relatively easy to remember.
6. **Bonus-level security:** Think of a sentence that means something to you (e.g., *I like to take long walks by the pier*), and take the first letter of each word to turn that into a combination of letters plus numbers and a symbol or two so it looks like this: ilttlwbtp2021$@. Although no password is technically "unhackable," this one comes pretty close.

GHOSTING: IDENTITY THEFT OF THE DEAD

In April 2020, the El Paso office of the FBI warned of scammers looking to take advantage of people with a deceased person in the family. The officials outlined a number of possible trolling motivations, including:

- **Outstanding debt:** False claims to the family of the deceased that the person owes money.
- **Funeral scams:** Offering funeral or memorial services that don't exist in an attempt to take money from people in a vulnerable state.
- **Medicare scams:** A scammer claims that Medicare is unpaid on behalf of the deceased and requests a payment.
- **Tax fraud:** Someone impersonates the IRS and claims that back taxes are unpaid.
- **Romance/compassion scams:** Less financial and more emotional, these scams may involve someone masquerading as the person who has died to either set up a fake relationship or perhaps even to acquire money for something like rent by impersonating them.
- **Delinquent life insurance ploys:** There are calls to grieving loved ones claiming that a life insurance policy was delinquent and asking for payments.
- **Credit card scams:** Opening false credit cards or lines of credit in a deceased person's name.
- **Specially engraved trinkets:** Thieves claim that a personalized item was intended for the surviving family member and request payment.

The New York State Department of State's Division of Consumer Protection labeled these kinds of identity theft attacks of the dead as "ghosting." There are a variety of nefarious means by which some are able to obtain logins, passwords, and other personal details to commit fraud:

- By cycling through made-up Social Security numbers in the hopes that one matches.
- By scouring places such as obituaries, funeral homes, hospitals, and other sites for personal details.

- By searching inactive social media accounts.
- By stealing death certificates. (When sending a copy of a death certificate, be sure to use certified mail with a "return receipt." This ensures that you know it's in the right hands.)
- By purchasing stolen items on the dark web (a deeply underground network of illicit activity often opaque to the authorities).

Once the attacker gains access to personal information, they can essentially pose as the deceased person, which means they can gain access to finances or money owed, or file false tax returns in order to get a refund. It also means they can open up dozens of credit cards with sizable limits in the name of the deceased person, and, as fast as possible, charge as many items as they can to those credit cards before the family or friends of the deceased realize what's happening. Attackers know this is a soft target with sizable returns, because the numbers add up: in 2011, the IRS reported that criminals reporting tax returns with names of the deceased were about to receive $5.2 billion.

It all happens during the weeks or even months between when someone dies and when formal institutions like banks or credit bureaus or even the Social Security Administration are alerted to the person's passing. The time window is wider since, most likely, no one is monitoring their accounts anymore, giving an attacker greater opportunity to bilk them out of more money. Then they skip town or disappear online, and the process to seek accountability is tedious if not impossible, especially when compounded by the grieving process.

Tech **TIP**

Although companies have a responsibility to protect your information (meaning, not to sell it or use it in ways that aren't clearly outlined in their policies), data can fall into the wrong hands. Only share information that you'd be comfortable with going public.

Tech **TIP**
Save important digital content to a local hard drive or to cloud storage. Things happen. Sites go down. Computers malfunction. Take ownership of your content and back it up to places that you control.

THIEVES PREY ON PANIC AND PANDEMONIUM

During the COVID-19 pandemic, the IRS paid out more than $207 billion in coronavirus relief payments to millions of Americans—some of whom had already died. (In some cases, the money was sent to people who had died more than two years earlier and whose deaths had been reported to authorities like the Veterans Administration or Social Security Administration, according to NPR.) Sometimes, that money became inaccessible to the remaining family members. Other times, it was transferred directly to a bank account that was still accessible by survivors. This presented them with a moral dilemma: Should they accept the money to help pay the bills, donate it to charity, or send it back?

Unfortunately, scammers thrive during this kind of confusion. They go to work in the wake of national disasters and pandemics as reporting agencies, hospitals, funeral homes, and other services become overwhelmed by a steady rate of reported deaths. The sheer volume of claims can overwhelm many authorities and the public, especially grieving individuals in need, who find it difficult to parse fact from fiction.

SYNTHETIC ID THEFT

There is also a kind of mounting techno-fraud that is part of a trend referred to as "synthetic ID theft." Perpetrators combine a mix of real and fake information to create a new profile of someone. For example, the profile could involve a phony name coupled with a real Social Security number (even from a child) and a legitimate address they've collected. The attackers can then establish a fake credit history over time until banks or other institutions are willing to loan them money. Once the credit lines are large enough, the criminals then "bust out" and convert the loans into cash or other convertible goods and disappear. They become your digital doppelganger.

Synthetic ID theft can be a serious issue since it's both hard to spot and challenging to correct because it contains elements of "real people" and facts. In 2016, synthetic ID theft was believed to be responsible for more than $6 billion in losses, which accounts for about 20 percent of credit losses, according to financial services consulting company Auriemma Consulting Group. And Experian estimates that the average loss related to synthetic ID theft is $6,000 with an increase of 35 percent from 2015 to 2016.

Tech TIP: Password Lockers

Imagine never having to remember multiple passwords! Password lockers, such as 1Password or LastPass, enable you to store many passwords in one place. A single unifying password can unlock all of the logins and passwords for your sites, like a locker containing all the relevant documents. It may sound risky to store all your passwords in one place, but, remember, that's all part of your online pact of trust with any company. Just be sure to keep that one important password—the one to your password locker—safe and memorized.

OUTSMARTING IDENTITY THIEVES

Whatever the reason for the unauthorized access to the deceased's account, there are things you can do to help minimize any chance of being a victim of a bereavement scam:

- Avoid putting too much personal information into an obituary like middle name, maiden name, exact birth date, address, and so on. (And be aware of how easy it is to spoof or alter an obituary, if it's required by a site for identification purposes.)
- Take action with social media accounts and other online sites (you may need legal advice here; see Chapter 10).

- Send copies of the death certificate to credit-reporting agencies as soon as possible. Don't wait for the Social Security Administration to do it (that process could take months).
- Review the deceased person's credit report for unusual activity, and ensure you have access to it beforehand.
- Send the IRS a copy of the death certificate.
- Contact the Department of Motor Vehicles to advise of the person's demise and to prevent duplicate identification from being issued.
- Contact the Direct Marketing Association to put the deceased individual on a "do not contact" list.
- Notify banks, credit agencies, and other financial institutions of the individual's passing.
- Speak only with trusted entities by contacting them directly yourself; don't entertain any incoming calls.
- Authorize trusted friends and family members, or "deputies," to shut down online accounts as soon as possible after your death. Or, at least, have them monitor the accounts so that they might receive any notifications from a bank or site asking for verification should those accounts be accessed.

NOT ALL BREACHES ARE FINANCIALLY MOTIVATED

When it comes to deception and the dead, not all scammers or hackers are looking for a payoff. Sometimes fraud involves pretending to be someone else just for the fun of it, the desire to take an inactive account on a veritable joyride. Such was the case in 2016 when the late *New York Times* media columnist David Carr had his verified Twitter account taken over by a spambot (Carr died in 2015). The account name changed to *Miranda Davis* and tweeted such exclamations as "I love role-playing games and sex." The breach only lasted for a matter of hours, but it was disturbing enough to affect many thousands of his followers, and others also reported receiving spam e-mails from his former account. Arguably, the last thing grieving families want is their loved one's name dragged through the mud.

Other times, the breach is a result of digital or human error. In one case reported by NJ.com in August 2020, sixty-two-year-old Margaret Tretola tried to buy a new dryer at a local Lowe's home center and applied for the corporate credit card to receive a 20 percent discount. She was declined (she ended up buying the item using her daughter's store credit card). Confused, because she had never been declined a credit card before and had impeccable credit, Tretola found out ten days later in a letter that the credit card bureau SageStream had her listed as deceased. Turns out, Tretola's Social Security number had been incorrectly added into its database.

A WORD ABOUT CRYPTOCURRENCIES

Cryptocurrencies are different from digital assets such as books, songs, or movies, as the latter are only available as part of a lifetime lease. Cryptocurrencies are essentially your money in digital form, but digital media are legally tethered to the account of the person who signs up and pays for them. In other words, when you buy an e-book from Amazon, you are just purchasing a license to read that book; you can't then *give* it to someone else.

CRYPTOCURRENCIES

Cryptocurrencies of any flavor—Bitcoin, Ethereum, Litecoin, and so on—are essentially digital assets that are part of a medium of exchange. These "coins" or currencies are stored in a ledger or computerized database, and strong cryptography (security) is used to verify ownership. The system does not require a central authority like a bank or country—thus, the term "decentralized"—and while a limited number of them can be used to purchase traditional goods and services or make donations to groups like the UN World Food Program, they are also used by criminal enterprises to try to hide funds since most of them are fungible.

The laws governing cryptocurrencies around the world are a bit of a mixed bag, and the jury is still out as to whether cryptocurrencies will one day replace all forms of centralized currencies. Still, they represent a sizable dollar value. Although all cryptocurrencies combined have a market of less than 0.7 percent of monies in the world, that number is still higher than $250 billion.

Yet, the IRS doesn't consider cryptocurrencies as currencies at all, but rather goods. At a state level, where enacted, the Revised Uniform Fiduciary Access to Digital Asset (RUFADA) Act establishes rules and regulations surrounding digital account ownership. RUFADA states that online management by a beneficiary (as in whomever you designate through the trading service) takes precedence over any will, estate trust, or power of attorney. However, access to the exchange does not equate to access to the funds. That's why planning ahead is critical to avoid any confusion.

The sharing of passwords or "recovery keys" is vital to accessing your cryptocurrencies upon your death. Without them, your money may be gone forever. Consider the case of Gerald Cotten, the founder of a cryptocurrency exchange called QuadrigaCX, Canada's largest cryptocurrency exchange. Cotten reportedly died unexpectedly at the age of thirty in December 2018 from complications of Crohn's disease while traveling in Jaipur, India. When that happened, more than US$190 million of cryptocurrency inventory or "cold storage" was lost in a black hole of inaccessibility. His widow, Jennifer Robinson, reportedly said that she was unable to recover the funds for investors.

In a court filing from 2019, Robinson wrote that while she had Cotten's laptop in her possession, she couldn't do anything with it. "The laptop computer from which Gerry carried out the companies' business is encrypted, and I do not know the password or recovery key. Despite repeated and diligent searches, I have not been able to find them written down anywhere."

In October 2020, the Ontario Securities Commission released a report that alleged massive fraud and misappropriation of funds on the part of Cotten and referred to it as a new-age Ponzi scheme that bilked 76,000 people of their investment. Some of the investors are unconvinced that Cotten is even dead and have formally requested that his body be exhumed and a postmortem report made public!

It's terrible to imagine that there are people out there who will take advantage of death, which should be off-limits for even the most ardent and vile scammers. Or that our digital assets can disappear entirely with our passing, or at the erroneous touch of a button. The good news is that you have the capacity to minimize your chances of being a ghosting victim or a victim of happenstance by taking a few simple steps to protect your online life. Once these vital layers of protection are established, then it's up to you to decide what you leave behind in the virtual world.

THREE KEY TAKEAWAYS

- Scammers are sophisticated and unscrupulous, and will seize any opportunity to take advantage of those who are grieving, because they are in a vulnerable state. It pays to be vigilant.
- Your passwords are the keys to your digital life and afterlife. Ensure they receive the same personalized protection as the keys to your physical home. Taking the time to consider their security will save heartache later.
- Outsmarting identity thieves takes planning and diligence, but is worth the time and effort. Alert companies and institutions as soon as possible on behalf of a deceased friend or family member to minimize the chance of being victimized.

CHAPTER 7

~~~~~~~~

Virtual Immortality
Memorializing Our Digital Worlds

"I thought I could describe a state; make a map of sorrow. Sorrow, however, turns out not to be a state but a process."

—C. S. Lewis, author of the Narnia series
Mere Christianity

When Fernanda Santos, a former *New York Times* reporter and now professor of journalism at Arizona State University's Walter Cronkite School of Journalism, lost her husband, Mike, to pancreatic cancer in 2017, their years of memories together, often chronicled in photos and social media posts, were priceless treasures stored in his iPhone: a time capsule of their love and commitment to each other. Santos held onto the phone and to the number, paying $5 a month to make sure it wouldn't be assigned to someone else, so she could give it to her daughter, Flora, when the time was right.

Isn't that what happens when someone we love dies? We try to hold on to whatever pieces they've left behind for as long as we can. We are reluctant to get rid of any belonging that might honor that person's memory or allow us to relive the happier times we shared with them. They serve as memorials to their life's journey.

PHYSICAL VS. DIGITAL MEMORIALIZATION

Physical Memorialization	
Pros	**Cons**
• **Tangible:** There's nothing quite like pulling out a family heirloom and holding it in your hands to remind you of a person or a memory, or walking into a gravesite to pay respects. • **Valuable:** If you've ever watched *Antiques Roadshow* on PBS, then you know that family heirlooms can have tremendous financial worth. • **Memorable:** The actual object holds the story or value, and its essence simply can't be captured in any digital fashion; visiting a final resting place generates a true occasion.	• **Vulnerable:** Even the most durable artifacts or memorials can be susceptible to the elements and vandalism. Monuments rise and fall. • **Losable:** If only one person knows the whereabouts of a family heirloom, it runs the risk of being misplaced or forgotten. • **Sizable:** Not everyone wants boxes of old family heirlooms in their basement or closet.
Digital Memorialization	
Pros	**Cons**
• **Affordable:** In many cases, minimal costs are associated with an entry-level memorialization. • **Accessible:** Anyone in the family or wider network can appreciate a virtual gravesite or photos of family heirlooms at any time. • **Available:** No need to travel to a gravesite or tomb to pay your respects. • **Customizable:** Filled with as much detail as desired. • **Immutable:** At least for the foreseeable future (see below).	• **Impermanent:** Digital decay means while the digital content is immutable for now, the site it's hosted on may go out of business, or the accessibility could disappear over a long period of time if the code changes or it can't be viewed with certain browsers or apps. • **Susceptible:** Hackers or online vandals can potentially deface sites or tributes. • **Antithetical:** For some people, the idea of not having a traditional memorial or preserving family heirlooms in a physical way goes against their beliefs or their religion and tradition.

MEMORIALIZATION OVER THE YEARS

Memorialization of loved ones was once the domain of the higher level of society. There was a steep cost associated with any meaningful preservation of someone's life: physical monuments (whose Latin origin is *monere*, which means "to remind") meant securing a plot of land, and burials weren't always possible depending on the location or the circumstances.

Early modest tributes to humans were constructed with wood or bone or stone. Over time, the more extravagant ones were designed with longevity in mind, such as the Great Pyramids, the Colossus of Rhodes, or the Taj Mahal. They aim to commemorate the life of the person and highlight their achievements and contributions while serving as a reminder to help mitigate the grieving process. (In more high-profile cases, they also serve as a reminder of their rule or their religious fervor or artistic accomplishments.)

As the concept of memorialization spread to the masses, the methods of paying respects also shifted to modest acknowledgments like creating gravestones or placing flowers on a gravesite. During the Victorian era (1837–1901), there was greater emphasis placed on memorials and the adding of inscriptions. The 1918 flu pandemic and other diseases along with multiple wars also created a need for graves and grieving on a mass scale.

Fast forward to the late 1990s, which represented a digital shift in the way we memorialized. When Princess Diana died in 1997, it was perhaps the first public moment when people, en masse, paid their respects online. The Internet was in its nascent stages then, and there was of course no social media, but according to a CNN.com article dated September 6, 1997, there were chat rooms established on the likes of Yahoo!, including a special section where anyone could leave their condolences and sign a memorial book. The Royal Network Online Memorial Service offered a guest book for visitors to sign before entering the site's "cyber chapel" that contained a site with Princess Diana's biography in images.

Sites started popping up to allow for online memorials and collective grieving. On a few sites, people even offered digitized songs in honor of her memory for download. The global movement even went so far as to generate online petitions and protests— including one called "Stop the Paparazzi!"—to criticize laws that failed to protect celebrities from prying photographers.

By 2001, the Internet had become more mainstream, and when the terror attacks of 9/11 occurred, they generated a massive outpouring of support online, further demonstrating the connectedness of unified grief. Some of those tributes remain online today. Then came the digital memorialization game changer: social media.

Did You Know . . . ?

In some places in Japan, when you visit a gravesite you can scan a QR code on the headstone that will play the favorite song of the deceased. Just another reason to consider your Spotify account as a window into your life to those left behind.

DEMOCRATIZING MEMORIALIZATION

Social media companies are by the far the most popular place to set up a memorial. According to a UK YouGov survey from late 2018, 7 percent of respondents wanted their social media accounts to be preserved after they die. That may not seem like a lot, but if you consider that there are more than one billion active users on Facebook, that figure alone represents about seventy million people.

Virtually every social media site or other digital service has its own rules for memorializing a profile. What follows are details for some of the most popular. Remember, it doesn't take long to check a few boxes— at the very least, it's better than doing nothing. If a digital service or app doesn't make it clear how to complete this kind of process, then contact customer service to learn more. Don't take no for an answer.

FACEBOOK AND INSTAGRAM

When Facebook started, it would routinely delete the accounts of people who had died. It was only after the Virginia Tech mass shooting in 2007 that led to the inclusion of memorialization of someone's account by 2009, following an outpouring of pressure from students and others.

That option still required users to submit a special form to convert the account into a memorial site where tributes could be paid or posts shared. By early 2015, it was possible to appoint a "legacy contact" for your Facebook account that empowered that person to manage your rights in the event of death. It also allowed Facebook users to have an option to delete their account when they die. (According to the Digital Legacy Association's 2017 study, 82.5 percent of respondents had neither heard of or pursued Facebook's Legacy Contact feature.)

Additionally, Facebook allows people to report the death of someone and adjust their account accordingly if they can produce an official form of documentation like a death certificate. The profile is altered such that status updates are disabled and contact information is removed, but loved ones can still post their memorials on the page in tribute to the account owner. (Unfortunately, in today's fast-paced world, social media posts are all too often the place where we first learn of someone's passing.) Eventually, the profile can be taken down if that's preferred by the friends and family of the deceased, and if that decision has been made in advance. Facebook advises users to memorialize the accounts of deceased loved ones as soon as possible so they can be secured.

With Instagram, a memorialized account is also hidden from public view, and no one can log into the account. Everything remains posted the way it was, and users can still send photos and videos through Instagram Direct. But unlike Facebook, there is no legacy contact option, and no portion of the profile is changed in any way aside from the word "Remembering" juxtaposed with the person's name.

TWITTER

As of early 2019, Twitter made it possible for users to deactivate an account, though previously posted content remains online. In order to deactivate an account, someone must produce a copy of the deceased person's ID along with a death certificate; if a friend or family member has the login and password of a Twitter account, then they are allowed to continue posting on behalf of the user (which can, of course, be slightly confusing or alarming to anyone not familiar with what happened). There is also some controversy around how much time needs to elapse before the handle of a deceased person on Twitter can be turned over to someone else.

LINKEDIN

LinkedIn has a page to contact the site and advise them of someone's passing by providing relevant information, such as their name, a link to an obituary (if it exists), and your relationship to the deceased person:

> We're sorry for your loss, and we appreciate your help in filling out this form.
>
> We'll use this information to make sure we're removing the correct profile.
>
> Thanks for your understanding.

The actions are then undertaken on the back end with a LinkedIn representative to act on the deceased's behalf by attempting to contact the account user.

PINTEREST

Pinterest is similar to LinkedIn and requires a copy of the death certificate to take action.

DROPBOX

Dropbox, the cloud storage site, reviews online activity, file shares, and overall engagement with the account over an ongoing period of twelve months to determine if perhaps the user has died. If an account is determined to be inactive, then it's deactivated. To gain access to an account of a deceased person, the heirs must submit proper documentation via physical mail or login if they know the credentials.

WORDPRESS

Wordpress is also similar to LinkedIn, and in addition to a copy of the death certificate, the website requires power of attorney or a legal document with a notarized statement to act on behalf of the deceased the actions to be undertaken with the account (e.g., gain access, shut it down).

THE FUTURE OF MEMORIALIZATION

Whether or not you want your body to be cremated, physically buried, or preserved for science, how you will be remembered online has become just as important. We have become a society that is used to having information and activities at our fingertips. Social media may be the memorial of choice at the moment, but others spring up depending on the day or circumstance.

For example, as the deaths related to COVID-19 rose in stark numbers in the spring of 2020, a couple in upstate South Carolina decided to create their own virtual memorial site dubbed "Mourning America" to bring a face to the numbers. The nonprofit effort offers rows and rows of photos of the people lost; clicking on the photos reveals greater detail about the individual and further it brings up a place to write comments or condolences. As the founders told a local news station, "This is a spot where people could come together and mourn loved ones that can't do it in the official way right now. We don't want anybody in this country to be limited from telling the story of their loved one."

There are even hybrids of real and virtual memorializations, such as the one offered by the company AFTR, which makes available an online camera view of anyone's gravesite to be viewed at their convenience on a laptop or mobile device. You can "visit" the person's final resting place anytime, an opportunity that became necessary during the COVID-19 pandemic as funeral services were limited, discouraged, or unavailable.

We know from history that offering the public a "place" to grieve or pay respects is essential—whether Lenin's tomb or Evita's grave or Elvis's home—and deleting them is tantamount to destroying their legacy. In late 2019 when Twitter issued a statement that said it would delete any "inactive" accounts, the reaction was swift and negative from many users. There was a desire to save those accounts as a place to memorialize anyone, not just celebrities and people with social standing. Twitter eventually apologized and backtracked. Facebook now offered a Tributes section, and even LinkedIn has a memorialization process under way.

In the future, digital memorialization may be through an extensive social media portal focused exclusively on obituaries and what we might learn from those who have passed, or it could even be through a VR experience like with Project Elysium. In 2015, the two cofounders announced an effort to capture the digital likeness of someone and allow others to interact with them in an immersive fashion after they died. Their words, their mannerisms, and their individual characteristics could be replicated and appreciated anytime (while wearing the right VR headset). (For more on the future of digital afterlife, see Chapter 10.)

MEMORIES ARE STORIES

Memories are stories. And stories are what makes us human. There is value in the preservation and transfer of them. What we learn helps us evolve over time—whether within a family dynamic or across a longitudinal social scale. And the medium to preserve those stories, the Internet, has made them easier to spread across a broader population, but also more interactive, more comprehensive, and more impactful.

We grieve together, and thus we grow together. And that opportunity and trend will only continue with new technologies and deeper understanding of the value to be gained.

But even the staunchest archivists among us admit there is some stuff—physical and digital—that we may not want saved or shared. And therein lies some questions to which only you have the answer.

THREE KEY TAKEAWAYS

- Memorialization has evolved from the physical to the digital as the Internet plays a more pivotal role in our lives.
- Each website has its own rules for how we memorialize our loved ones. Make sure you know the options available, especially if they're public in nature.
- Digital memorialization continues to grow as companies as well as individuals seek options for how we both celebrate and grieve those who have died.

CHAPTER 8

When It's Better Not to Remember
Shut Down and Delete

"Sometimes you just have to erase the messages, delete the numbers, and move on."

—Nitya Prakash, Indian author

The concept of saving someone's digital life and sharing it with loved ones has a profoundly noble ring to it. We see the value in passing on all that wisdom and experience and memories down to future generations, and saving what's meaningful: photos, music, personal memories. But the truth is that we may not want everything saved and shared. Our complicated lives were lived and shared online and in the moment without any thought to what we would leave behind, and it shows.

We leave inebriated paths of digital breadcrumbs all over the world, across servers and sites and devices, and not only do we not know where we're going or where we'll end up, but the people who try to follow us after death can be left puzzled and lost without any sense of even where to begin. We have two choices:

- **Aggregate approach:** Leave everything or delete everything. Or leave that decision up to your deputies. This is the "all or nothing" approach that leaves you with minimal responsibility for now but shifts the burden to those left behind to try to interpret your wishes.
- **Selective approach:** Keep certain digital slices of your life while eliminating others. Save the photos in the family album but

delete the ones in your dating profile. Allow access to Facebook, but not Twitter. This takes more work on the front end, but allows for greater peace of mind for everyone involved.

We've spent much of this book trying to figure out how to protect our digital afterlife and make it accessible to those we love. But what about the stuff that we don't want available? The stuff that we'd rather people didn't see? There may be quite a bit:

- **Credit card bills and debt:** Embarrassing, and it usually comes to light somehow through collectors.
- **Time spent in online games (and the money spent there):** Bills will keep coming.
- **Goofy videos watched on YouTube:** Some might be shared publicly—you may also have favorite videos saved.
- **Your general web-surfing habits, including porn (no judgment):** Discoverable in your browsing history.
- **Online gambling:** Also discoverable through your browsing history or bill statements.
- **Photos from ex-relationships:** Perhaps stored, perhaps shared.
- **Outtakes from selfies:** Yes, those are out there somewhere.
- Text messages: Having your text messages available is bad enough, but having them out there *without context* is worse. If someone has access to your phone, they're readily accessible unless they've been hidden with an app like CoverMe, Wire, or Signal.

There's a combination of the internal stuff we do and save (documents, e-mails, financials, personal photos, videos) and the external stuff we share and post (public photos, social media posts, work history), and shutting them down or deleting becomes a game of Internet search and destroy.

Tech **TIP**

Close the accounts you no longer need—while you're still alive. Not only will it have a calming effect on your mind by eliminating the digital clutter, but it can also cut down on any chance you could face fraud or scammers in the future who gain control of unused accounts and try to impersonate you. Rule of thumb: If you aren't using an account at least on a monthly basis, then you probably don't need it.

GOOGLE

According to the Digital Legacy Association's 2017 study, almost 90 percent of respondents had either not heard of Google's Inactive Account Manager or hadn't pursued it, and it's easily one of the most robust options. It's an algorithm designed to see if you haven't logged in recently or used any Google-powered device (Android). You can set the inactivity parameters based on personal preference: for example, you can set the inactivity range to be anywhere from three months to eighteen months and assign a trusted contact to receive a note that might look like this:

John Doe (john.doe@gmail.com) instructed Google to send you this mail automatically after John stopped using his account.

Sincerely,

The Google Accounts Team

The person you've assigned—only one—will then have access to whatever information you've designated for them to access for three months, including Google products like Maps, YouTube, Gmail, and so on. A follow-up e-mail to that trusted person might look like this:

John Doe (john.doe@gmail.com) instructed Google to send you this mail automatically after John stopped using his account.

John Doe has given you access to the following account data:

Blogger

Drive

Mail

YouTube

Download John's data here.

Sincerely,

The Google Accounts Team

You can also choose to have Google delete everything—photos and documents—that have been saved. Also, YouTube, which is owned by Google, grants access to certain people under the requisite circumstances (the Inactive Account Manager option also covers YouTube).

HEY, GOOGLE: DOES DELETED MEAN GONE?

Google (and Facebook) states that on request, after a certain period of time (between thirty and ninety days), all user data will be permanently deleted if you choose to delete an account:

> You're trying to delete your Google Account, which provides access to various Google services. You'll no longer be able to use any of those services, and your account and data will be lost.

When in doubt, follow up with Google (or any company) to get verification.

DATING SITES: A SEARCH FOR LIFE PARTNERS IN THE AFTERLIFE

The provenance of dating for our parents and grandparents is a little blurry. Maybe they met at the local diner; maybe they were both in a dance club together or spotted each other in a crowd. One thing we know for sure, though: however they met, they didn't assemble all their traits and attributes—and some of their most intimate details—into a portal for others to see and browse.

But *we* do. Online dating has become one of the most popular ways couples meet. According to a 2017 survey of American adults,

39 percent of heterosexual couples reported meeting their partner online, compared to 22 percent in 2009. How much of that digital information would we want to save for future generations? Or maybe a better question is: How do we go about getting rid of those embarrassing come-hither photos?

Let's use Match.com as an example. According to Nolo.com, a self-help legal guide designed for consumers and small businesses, the most important consideration for anyone with a Match.com profile or other online dating site is to leave clear instructions for your deputy, in a will or other legal document, for deactivation or deletion in the event of your passing.

For example, you may wish to have the profile deleted or deactivated. Simply canceling a subscription will not remove that profile from the site. Even if you delete or deactivate your profile, that information may not be entirely gone or private as Match.com may keep it for up to a year in case the profile is reactivated.

Plus, other members may have saved certain photos or messages from you within the platform. Unless those users are willing to delete them, they may be preserved over time. Another reason to consider what you share about yourself online.

Tech **TIP**

Make sure legacy lists are current. Just as wills are updated as life circumstances change, so should legacy lists. We may have tasked our deputies with shutting down our social media accounts upon our death, but if we've changed the passwords or login information to those accounts, as we should periodically, then it's like handing them the wrong key to a lock.

REGISTERED DOMAINS

Many of us have registered domains for businesses or personal reasons, such as wedding registries. To shut them down after you die, task your deputies with logging into the site where the domains are registered and clicking on the account settings (e.g., with GoDaddy, this area is called the Domain Control Center). There, you can choose which domains to delete. Note: If there is ownership protection, you may need to verify your identity through a two-factor authentication code that's sent to a corresponding device via text message, or through a one-time password sent to a corresponding e-mail address.

SUBSCRIPTION ACCOUNTS

Subscription accounts range from news portals to Netflix and Blue Apron. Although each has its own accessible process for canceling that account or stopping the billing cycle, fully deleting an account may require contacting customer support, although Netflix will automatically delete a canceled account after ten months.

BANKING/PAYMENT SITES

If a bank account is a custodial account that names you as the pay-on-death beneficiary, then you must request a certified copy of the death certificate from the state's office of vital records and present it to the bank with identification. The bank should then release the money to you and allow you to close the account. With payment sites like PayPal you will need to access the site's settings and choose Account Options and Close Account. You may need to enter some bank information as verification of identity. Any unpaid money requests are automatically canceled.

CLOUD-BASED STORAGE

Although Google Drive can be deleted (along with all the contents: photos, documents, videos, etc.) through the Inactive Account Manager (see page 80) or within your Google account profile, with a cloud-based storage site like Dropbox you would need to log in and, under Settings, choose General and Delete account. Files stored locally on your computer in a Dropbox folder will remain, but any files stored on Dropbox servers will be deleted.

MESSAGING APPS/TEXT MESSAGES

Messages on apps such as WhatsApp can be deleted by selecting the conversation and swiping. However, that only deletes those messages on that device and not the other person's. A "Delete for Everyone" feature is only available within an hour of the messages being sent and is therefore limited in scope. Similarly, basic text messages can be deleted by selecting them on a device. With Apple Messages, for example, you can choose to delete a message or an entire conversation; if iCloud is turned on, that will also delete them across other devices.

AMAZON

When deleting an Amazon account without, if you do not have access to your loved one's login information, you can close the account through Amazon's customer service chat or by e-mailing Amazon customer support. If you have your loved one's login information, simply access Amazon's Help page to submit a request to delete the account. Then log in and choose "Close my account and delete my data" in the dropdown menu, and follow the instructions Amazon provides.

GAMING SITES

With Google Android apps, you can delete an account or data tied to a specific game within the Play Games app. By clicking on Settings, you can choose to delete play games account and data or choose to delete individual game data. A similar process is available within Apple iOS by clicking on your Apple ID Profile and iCloud. Then click on Manage Storage, and delete the game and data you choose. With a gaming service like Epic Games, you would need to log into the profile and choose General Info to select *Delete Account*: a security code is then sent to the account holder's e-mail address to confirm deletion. In the case of poker sites, such as PokerStars, shutting down an account may involve sending an e-mail to customer support to formally request that the account be closed and deleted.

BROWSING HISTORIES

In this case, it really depends on the browser you're using:

- **Google Chrome:** Click the *History* menu at the top of the screen, choose *Show Full History*, and then select *Clear browsing data*.
- **Firefox:** Choose *History* and click *Clear Recent History* (select *Browsing* and download *History* as well as *Form & Search History*).
- **Internet Explorer/Microsoft Edge:** Click the three bars on the right side of the web address field and click on *History*. Then click *Clear History* at the top (and ensure *Browsing History* is checked).

Keep in mind that computer forensics experts can likely still find data you want really hidden, but these methods will at least remove anything readily accessible.

DUE DILIGENCE

Our digital lives are more scattered than our real lives ever were. With physical objects—car, house, clothing—we see them and generally know where they were or have some rough mental inventory of them. Online, however, we don't exactly know where the stuff we create every hour of every day—social media posts, Google docs—is stored or who even has access to it. Take the time to find out, and do your due diligence. The process may seem time consuming, particularly if you have a robust online life, but it's worth taking control of whatever information you can because, as the next chapter discusses, your rights may not extend to everything with your name on it.

THREE KEY TAKEAWAYS

- Not everything we do online is worth keeping—or remembering. Sometimes there are things you want to delete and want to preserve within the same platform! Consider whether you want to shut down or delete your information in the aggregate (with a giant ax) or selectively (with a scalpel).

- Take advantage of website features such as Google's Inactive Account Manager that will alert deputies and trusted contacts when there is no activity from you on that site.
- Think about every website you visit that you take great pains to make sure no one knows about. Now's the time to make preparations to make sure no one knows about them after you're gone as well.

CHAPTER 9

What Are My Rights?

Privacy and Security Laws

"Our own information is being weaponized against us with military efficiency. Every day, billions of dollars change hands and countless decisions are made on the basis of our likes and dislikes, our friends and families, our relationships and conversations, our wishes and fears, our hopes and dreams. These scraps of data, each one harmless enough on its own, are carefully assembled, synthesized, traded and sold."

—Tim Cook, CEO of Apple

Throughout the evolution of technology and manufacturing, laws often lag behind innovation and invention. When cars became mainstream, it took the wonky nature of activist and politician Ralph Nader and an arsenal of data to convince lawmakers that seatbelts were not an inconvenience but a necessity. When MP3s and digital music became wildly popular through ripping songs and sharing them online in the early 2000s, it took years before the act was deemed illegal. And eventually it took another tech company, Apple, to offer a better product—a broad range of AAC music files that worked seamlessly within Apple devices and only cost $0.99—to help push people to select a quality and legitimate option.

When it comes to information, rather than invention, the legal field gets a little muddied. For example, the Twitter account of Republican politician and former presidential candidate Hermain Cain remained active after his death in late July 2020. The account was maintained by members of his team and family, and even continued to react to the political events of the day; the verified account was eventually changed to the moniker *The Cain Gang*, which allowed it to adhere to Twitter's rules of who is managing the account.

Therein lies another dilemma: when no instructions are given, who should make the decisions about memorialization. Should high-profile accounts be deactivated or memorialized as a testament to history? Institutions such as the Library of Congress and the Internet Archive preserve selective tweets for posterity. Shouldn't the average person have a clear understanding of whether they're reading something that was directed by the deceased or generated by their remaining circle of peers and loved ones? Or even created by a bot?

We've talked a lot in this book about how important it is to assign deputies and make preparations regarding your digital assets and data for when you die. Even so, there are also virtually *no* official provisions to pass your online data on to others in a will or trust—*if* there is a legal challenge. In other words, you might wish a friend to delete photos from your Facebook account upon your death, but your spouse might challenge that in court.

WHO DECIDES WHAT?

A case in France in September 2020 shed light on an end-of-life dilemma for tech companies as a man named Alain Cocq was denied the chance to stream his last moment of life by Facebook and the French government. Cocq, fifty-seven, said he had suffered for more than thirty-four years from a degenerative and painful illness and appealed to President Macron to die with dignity in this public manner (which, of course, could be recorded by anyone and potentially preserved as part of his legacy). Macron offered his sympathies but did not overrule the decision by Facebook to block the livestream.

"The path to my deliverance is starting and, believe me, I am happy about it," Cocq said at the time he stopped taking any fluids, according to the Associated Press. "To those I won't see again, I say goodbye. Such is life."

There is no legal onus for government or tech companies or any companies with a digital presence to do what's "right" or what's preferred by family members or friends, residents, and neighbors.

Consider the case of a fifteen-year-old German girl who died in 2012 after being hit by a subway train in Berlin. Her parents wanted access to her Facebook account, but in spite of having the passwords, it had been memorialized, and they couldn't access her account to help confirm whether her death had been a suicide or an accident. Eventually, they sued Facebook for access, and while they lost that case, the German courts ultimately granted permission to get into her account—albeit nearly six years after her death. Even with newer laws around privacy, such as the U.S. Data Protection Act of 2018 or General Data Privacy Rules (GDPR) in Europe (see page 84), they rarely if ever apply to the deceased.

Did You Know . . .?

In order to protect user privacy, the Stored Communications Act in the United States prevents tech companies from readily turning over personal information. But what if you live in another country and need access to personal information from a tech company based in the United States? Or vice versa? In 2018, Congress passed the Clarifying Lawful Overseas Use of Data Act, which gives American law enforcement greater powers to access U.S. citizens' accounts held on overseas servers. But it doesn't apply to every country, and the reciprocal nature is often vague, nonexistent, or wrapped in a protracted legal maze.

GEOGRAPHICAL HURDLES

Then there's the issue of geography. Tech companies may have a headquarters in Silicon Valley or New York or Beijing or London or

Sydney, but the users are scattered across countries, and the families of those people may also live on different sides of a border that are governed by different laws.

- **Europe:** Enacted in May 2018, the General Data Privacy Rules (GDPR) extends strict jurisdiction over the protection of personal data in Europe, and even beyond European countries to cover any global business that sells to or has EU customers. It covers everything from storage of data to defining a clear purpose of its usage to security and integrity of the information. It also applies to the transfer of data across borders. In other words, even a U.S.-based company such as Google needs to adhere to GDPR when it comes to any EU customers to ensure their privacy is protected. Even with Brexit, the United Kingdom is expected to still comply with the tenets of the GDPR, though it will become known as the UK GDPR.

- **United States:** The U.S. Uniform Law Commission, an association empowered to suggest laws to be enacted across the country, introduced a law in 2014 and 2015 that was intended to allow access to any online accounts of the deceased (no matter the age of the person involved) by the loved ones unless otherwise stipulated in a will or trust. But even after the bill was proposed in twenty-six state legislatures, tech companies like Google, Facebook, and others pushed back with an alternative version that requires people to obtain court orders to gain access to someone's account after they die. As anyone who has tried to get a court order knows, that can be a time-consuming and tedious process, especially for those also dealing with loss. A watered-down version was adopted in forty-four states but still makes it relatively difficult for anyone to get into someone's account, even if it's a close family member, such as a parent trying to learn more about what happened to their child.

- **California:** Originally drafted in 2018, the California Privacy Rights Act went into effect in early 2020 as a model to the rest of the United States. It is designed to expand the rights of consumers to (1) know more about how their personal data is collected and used, and (2) retain the right to have it deleted and opt out of it being monetized. But between 2018 and 2020, some tech companies and lobbyists discovered loopholes and sought to dilute it. Although the act has high-profile proponents like former presidential candidate Andrew Yang and Common Sense Media, it also has unexpected detractors like the ACLU of Northern California, which claims the act will make the experience more onerous for consumers who opt out and may even force customers to "pay for privacy" through higher costs if they refuse collection of their data.

- **Canada:** In Canada, legal experts have attempted to mitigate any confusion with something called the Uniform Act on Fiduciary Access to Digital Assets (UAFAD). It stipulates that any virtual property should essentially be treated the same as any physical assets in being passed on to the respective heirs. In other words, it doesn't make a distinction between an old-fashioned filing cabinet and an online game account.

This patchwork of requirements and settings is not only confusing for many users, but it also means they are often forced to circumvent any restrictions by trying to obtain a user's login and password information to manage a loved one's digital afterlife themselves—without the involvement or approval of the tech company. Think of it as a slightly underground approach to digital afterlife management.

For example, when someone dies the family may want to alert contacts about what happened or delete the deceased's profile entirely, but either way they feel compelled to act as soon as possible to minimize any swirl or heartache that could result from misunderstandings or misuse. Why worry about the possibility of violating an obscure law or rule when a person's legacy is at risk of being exploited or trampled? Yet, all these good intentions may be violating a website's terms of service and setting up for legal headaches or risk.

WHEN TO SECURE LEGAL COUNSEL

Gaining access to accounts often depends on whether you are the next of kin or named in the will as a beneficiary or retain the power of attorney. But what if you're not named as a beneficiary and don't have access to the will but want to secure the photos of you and your fiancé from a social media site? That's certainly an instance when a lawyer can advocate on your behalf to at least inquire whether the deceased left logins and passwords in their will and make a case for receiving them. Proceed with caution if you want to try doing it on your own.

INTELLECTUAL PROPERTY BATTLES

As tech companies create new products, the intellectual property battles often drag on so long in the courts—like the one between Apple and Samsung over smartphone features that lasted seven years, from 2011 to 2018—that the original laws governing those actions become woefully inadequate and antiquated. On the flip side, if we expected the corporate world to wait for governments to catch up, then progress would slow to a crawl. The ability of tech companies to be agile and adapt to the marketplace doesn't comport with the machinations and pace of Congress or Parliament or any other elected body, so consumers are usually left somewhere in the middle. They are affected by both ends of the spectrum with ongoing concerns about their privacy, rights, and options.

Tech companies generally don't change their products unless and until their consumers make it clear they aren't happy and demand something better. Sure, there are always updates and additions and improvements, but shifting priorities is driven by a bottom line and something like managing anyone's digital afterlife is now at an inflection point. None of the tech companies have completely solved these challenges, and for now the approach is a patchwork of privacy considerations, legal ramifications, public demand, personal interests, ethical puzzles, and technological limitations. There must be more attention paid to this issue as the world's population continues to swell and move online at an exponential rate. Even with modest improvements and statutes to provide users with more rights, there is no standardization of policy globally. It's time that laws reflected reality when it comes to our privacy, our rights, and the rights of our family, while ensuring that tech companies are held responsible for addressing it all in a humane, thoughtful, and dignified manner. We owe it to future generations to create better policy and management to respect the dead and provide them a true state of R.I.P.

THREE KEY TAKEAWAYS

- Knowing your rights sounds like a cliché—and it is. But it doesn't make it any less relevant or important especially when it comes to a relatively nascent space like the preservation of information after someone dies. Do your legal research, and don't be afraid to ask questions.
- We live in a globally connected world with a mishmash of privacy and security laws that are constantly playing catch-up. But that doesn't mean giving up. Source the experts who know the current situation; sometimes that means hiring legal counsel.
- The employees at tech companies are human, too, and, ideally, are sympathetic to what it means to obtain information about one's digital life. As we slide further along this time line of merging online and offline worlds, it's up to tech companies and others, including government officials, to help take the lead and ensure people are informed and empowered.

CHAPTER 10

The End (Beginning?)
A Constant State of Digital Evolution

"The connectivity of the cloud and the prevalence of tablets and smartphones have eroded the traditional online/offline divide. Within a short time we will most probably stop thinking of it as 'online.' We will simply be connected, all the time, everywhere, and the online world will be notable only by its absence when that connection breaks."

—David Amerland, author, *Google Semantic Search*

In the 2004 science fiction thriller *The Final Cut*, the late Robin Williams plays a character whose job is to craft a highlight reel of a person's entire life, culled from visual and audio footage collected from an implanted microchip. The reel will be viewed by friends and family and preserved over time, and he must decide what parts to leave in and take out. How should the person be remembered? What if he or she has committed both heinous acts of violence and charitable works? What is the sum of a human life? What moments define us? Some more than others? And who decides? These are important questions that, most of us would probably agree, should not be left to a third party to answer.

In some ways, many of us already have begun creating that highlight reel on social media, but there are also people embarking on a real-life version of this experiment.

Justin.tv was started in 2007 by Justin Kan. Kan began by broadcasting every moment of his existence—a real-life version of the 1998 film *The Truman Show* (although Truman was unaware he was the star of his own TV show and Kan was wearing the camera). Justin.tv eventually helped coin the term "life-casting." His experience led to an entire platform of people doing something similar (some still do). But this prompted additional questions related to everything from the broadcast of copyrighted material to ethical issues: for example, at least one teen used the Justin.tv platform to post a live act of suicide.

These days that life-casting concept has shifted slightly away from exposing every intimate detail—though there are still plenty of reality-TV options—to sites like Twitch.tv, which allows anyone to watch their digital gaming experiences and interact with them in real time. What else might the future bring?

Tech TALK

Internet of Things (IoT): Originally coined by British technology pioneer Kevin Ashton in 1999, Internet of Things refers to the concept that physical objects can be connected to each other with unique identifiers and transmit data with human mediation. By 2011, the term started to emerge as a growing sector and now collectively refers to things like connected light fixtures, voice control, and home appliances, but also broader areas such as smart agriculture, monitoring systems, and autonomous vehicles.

A FUTURE OF DIGITAL POSSIBILITIES

Imagine if tracing the roots of your ancestors meant not only putting together a family tree but also generating avatars of their likenesses that could channel the wisdom and experiences from their lives. What about layering in an Internet of Things (see box) aspect such that you could communicate with a photo or family heirloom and hear the stories tied to it from your ancestors? What if they could help you in the present through the wisdom gained from their experiences, even after they die?

In 2018, there were seven billion IoT devices, everything from lightbulbs to refrigerators to connected cars. By 2019, that number soared to 26.7 billion devices, and it's steadily climbing.

By feeding into an algorithm, we could resurrect anyone with the right amount of data and allow them to interact with future generations. The example of holographic performances from the likes of Prince at the Super Bowl in 2018 are one example; in recent years, other groups have taken on more socially minded applications like the ability to converse through augmented reality (AR) with holocaust survivors at the Illinois Holocaust Museum and Education Center. The organizers at the museum interviewed more than twenty holocaust survivors and then allowed anyone in the museum to ask their own questions to the hologrammic version. The more questions their database was able to sort, organize, and learn from, the better it could tether the right answer from the survivor.

The company behind this effort, StoryFile, is already offering anyone the chance to preserve their legacy with some amount of control. The process involves sitting down in front of a special camera setup to record answers to a series of questions about values, life lessons, personal struggles, morals, decisions, and more to inform a profile of the person through AR/VR. How many of us wanted to sit our parents down in front of a camera and ask them a series of questions to save for their grandchildren? It's like a time capsule of their life—and more comprehensive than a cluster of social media posts, and richer than various photos and random videos.

In 2019, a mother in South Korea was able to "see" her deceased daughter through a VR experience that combined AI, voice recognition, and 3D capture. The heart-wrenching moment was captured on video, thanks to a documentary called *I Met You* and broadcast to a huge audience across the country. The mother's grief is clearly real and uncomfortable to watch, but she seems somewhat comforted by the experience to see her daughter "come back to life."

In 2017, Toronto-based grief counselor and thanatologist (the study of death and its impact on others) Andrea Warnick told the website Quartz that the creation of chatbots that replicate a deceased loved one could play a role in not only simulating an interaction but facilitating important conversations within the network of the bereaved.

"In modern society, many people are hesitant to talk about someone who has died for fear of upsetting those who are grieving," Warnick told Quartz, "so perhaps the importance of continuing to share stories and advice from someone who has died is something that we humans can learn from chatbots."

Creepy or cool? Perhaps it depends on whether it's your family or someone else's—like in October 2020 when Kanye West gave a hologram to Kim Kardashian of her deceased father, Robert Kardashian. She loved it, but others weren't entirely sold on the idea. The bottom line is that technology has created this challenge, so there are—and will be—novel ways for technology to be part of the solution. What we choose will be up to each of us.

A TEAM EFFORT

But how far down this road do we want to travel? The ethical questions of digital afterlife seem endless: Are public or historical figures of greater interest to the world and therefore exempt from such considerations? Should families feel a need to limit the resurrection of their loved one(s) or an obligation to their fans and followers? How do we balance saving someone's legacy and sharing it? What happens to everything Mark Zuckerberg has ever created on Facebook? Should that answer be any different than that of any other user? The answers to these questions come from collaborative conversation with:

- **Top executives:** A top-down approach from the C-suite level to contend with liability issues, from a compliance and corporate level.
- **Lawyers:** To make sure companies are following tricky privacy laws.

- **Government:** To make sure the laws on the books are keeping pace with technology.
- **Employees:** A groundswell of employees to underscore the importance of this issue.
- **Engineering teams:** To create better user experiences or setting up surveys to understand the various pain points or eliminating silos between competing companies to allow for a universal interface and database.
- **Ethicists:** To address issues that may arise. As AI and machine learning grow exponentially in their ability to learn from data sets and infer correlations, maybe we're not happy with the final "version" of a deceased person. Maybe there are needs within certain ethnicities or demographics to infuse the right empathic language. Or we need to create an option to keep these high-tech experiences private to a closed group versus broadcasting them (as with the South Korean example). That way, a family could collectively choose to "reconnect" with someone they lost.
- **Grief counselors:** To facilitate the entire experience.
- **The public:** We can all become Narcissus if we don't hold ourselves accountable and occasionally step back from the reflective digital pond.

The largest digital companies have the opportunity to lead by example and pave the way for others to follow by setting standards and generating discussion. We need to lean heavily on science, the law, pragmatism, and philosophy to ensure we don't feature in our own episode of *Black Mirror* or cause more harm than good.

ACT WHILE YOU CAN

Very few of the choices surrounding our digital demise are straightforward or easy. Though that may change in the future, for now it's important to understand how much power you have and how much remains in the hands of others, particularly the companies that manage, and potentially own, all your information and data. Yet, there is no need to wait for Silicon Valley or the federal government or anyone anywhere to pull out our compass and move forward.

Consider all the personal stories, lessons learned, and experiences shared that are in the balance. On a grand scale, think about human evolution and development trends that could be valuable for historians and analysts. We have a duty not only as individuals but also as global citizens to get our digital affairs in order, and that means both protecting our access and turning on/off the access of others. It's also not about entirely replacing any analog means of preserving our human stories (e.g. oral sharing, journals, art, heirlooms, etc.) but rather ensuring that we don't lose our digital legacy along the way.

It's about control—who has it now, and who gets it after we die.

Right now, you have it. Use it well.

Resources

CHAPTER 1

https://www.statista.com/statistics/346167/facebook-global-dau

https://www.omnicoreagency.com/instagram-statistics

The Digital Afterlife Project: https://www.digitalafterlife.online

https://www.forbes.com/sites/bernardmarr/2018/05/21/how-much
-data-do-we-create-every-day-the-mind-blowing-stats
-everyone-should-read/#3602161360ba

https://techjury.net/blog/big-data-statistics/#gref

Digital Legacy Association: http://digitallegacyassociation.org

https://www.westmonroepartners.com/perspectives/point-of
-view/americas-relationship-with-subscription-services

https://digitallegacyassociation.org/wp-content/uploads/2018/07
/Digital-Death-Survey-2017-HQ.pdf

https://www.pewresearch.org/internet/fact-sheet/mobile

https://www.theloop.ca/dead-facebook-users-will-soon
-outnumber-the-living

CHAPTER 2

https://www.insider.com/how-computers-evolved-history-2019
-9#in-the-1940s-computers-took-up-entire-rooms-like-the
-eniac-which-was-once-called-a-mathematical-robot-2

https://www.ibm.com/ibm/history/ibm100/us/en/icons/tabulator

https://wearesocial.com/blog/2019/01/digital-2019-global
-internet-use-accelerates

https://makeawebsitehub.com/social-media-sites

https://www.broadbandsearch.net/blog/average-daily-time-on
-social-media

https://www.mediapost.com/publications/article/348093/gen-zs
-favorite-social-media-platform-none-of-th.html

https://techjury.net/blog/how-much-data-is-created-every
-day/#gref

https://www.psychologytoday.com/us/blog/all-about-sex/201611
/dueling-statistics-how-much-the-internet-is-porn

https://www.alliedmarketresearch.com/mobile-application
-market

https://www.mediapost.com/publications/article/291358/90-of
-todays-data-created-in-two-years.html

https://www.statista.com/statistics/746230/fortnite-players

https://www.businessofapps.com/data/roblox-statistics

https://deepmind.com/blog/article/alphazero-shedding-new-light
-grand-games-chess-shogi-and-go

https://analyticsindiamag.com/deepmind-vs-google-the-inner
-feud-between-two-tech-behemoths

CHAPTER 3

https://alltechishuman.org

https://morningconsult.com/wp-content/uploads/2020/01/Most
-Trusted-Brands-Executive-Summary.pdf

https://www.pwc.com/us/en/services/consulting/library
/consumer-intelligence-series/trusted-tech.html

https://www.varonis.com/blog/company-reputation-after-a-data
-breach

https://www.pkware.com/blog/what-s-the-real-cost-of-a-data
-breach#:~:text=Data%20breaches%20are%20simply%20
a,insiders%20on%20a%20daily%20basis

https://www.inc.com/larry-kim/22-things-big-tech-companies
-know-about-you.html

CHAPTER 4

https://mygoodtrust.com/store/site

https://ec82ed81-fdf3-475e-91c7-f4e5e324a9ff.filesusr.com
/ugd/10c674_f4a951fcff394fbdbbb5b10f8eef0cf6.pdf

CHAPTER 5

https://www.wsj.com/articles/how-to-prepare-your-financial
-information-for-when-you-die-11601697960

CHAPTER 6

https://purplesec.us/resources/cyber-security-statistics

https://www.dos.ny.gov/consumerprotection/scams/afterdeath
.html

https://www.rollcall.com/2019/12/10/regulators-warn-about
-fraudsters-creating-synthetic-borrowers

https://markets.businessinsider.com/news/stocks/synthetic
-identity-fraud-cost-banks-6-billion-in-2016-auriemma
-consulting-group-1002222563#

https://www.nj.com/news/2020/08/you-cant-have-any-credit
-because-you-are-listed-as-deceased-store-tells-customer.html

https://www.npr.org/2020/05/06/851019441/the-irs-sent
-coronavirus-relief-payments-to-dead-people

https://www.theguardian.com/technology/2019/feb/04
/quadrigacx-canada-cryptocurrency-exchange-locked-gerald
-cotten

CHAPTER 7

Mourning America: https://mourningamerica.org

https://slate.com/technology/2020/03/digital-afterlife-death
-social-media-iphone.html

https://www.vice.com/en/article/ppxay7/when-you-die-facebook
-decides-what-to-do-with-your-profile-720

https://www.aklander.co.uk/news/history-gravestones

http://www.cnn.com/WORLD/9709/06/diana.internet

https://yougov.co.uk/topics/lifestyle/articles-reports/2019/11/01
/what-do-brits-want-happen-their-data-and-social-me

https://digitallegacyassociation.org/wp-content/uploads/2018/07
/Digital-Death-Survey-2017-HQ.pdf

https://help.latest.instagram.com/231764660354188

https://digitallegacyassociation.org/wp-content/uploads/2018/07
/Digital-Death-Survey-2017-HQ.pdf

https://www.linkedin.com/help/linkedin/answer/2842/deceased
-linkedin-member?lang=en

https://support.google.com/accounts/answer/3036546?hl=en

https://www.linkedin.com/help/linkedin/answer/2842/deceased
-linkedin-member?lang=en

https://wordpress.com/support/deceased-user

https://www.everplans.com/articles/how-to-close-a-pinterest
-account-when-someone-dies

https://www.wtsp.com/article/news/health/coronavirus/digital
-memorial-honors-lives-lost-coronavirus-mourning-america
/67-aeb737d8-6988-4e01-8f85-f1fcaa4453a2

CHAPTER 8

https://www.investopedia.com/tech/how-much-worlds-money
-bitcoin/#:~:text=Bitcoin%20is%20the%20largest%20
and,comes%20to%20roughly%20%24251.8%20billion

https://www.osc.gov.on.ca/quadrigacxreport/index.html#executive
-summary

https://news.bitcoin.com/quadrigacx-founder-dead-or-alive
-request-for-exhumation-and-autopsy-filed

https://www.nolo.com/legal-encyclopedia/what-will-happen-my
-matchcom-account-when-i-die.html

https://www.facebook.com/help/356107851084108?ref=dp

https://news.stanford.edu/2019/08/21/online-dating-popular-way
-u-s-couples-meet

CHAPTER 9

https://www.theguardian.com/technology/2017/may/31/parents
-lose-appeal-access-dead-girl-facebook-account-berlin

https://www.secureworldexpo.com/industry-news/tim-cook
-privacy-quotes

https://www.extremetech.com/extreme/55210-napster-vs-itunes

https://www.washingtonpost.com/technology/2020/08/31
/herman-cain-twitter-account

https://www.theverge.com/2018/6/27/17510908/apple-samsung
-settle-patent-battle-over-copying-iphone

CHAPTER 10

https://qz.com/896207/death-technology-will-allow-grieving
-people-to-bring-back-their-loved-ones-from-the-dead-digitally

https://securitytoday.com/Articles/2020/01/13/The-IoT-Rundown
-for-2020.aspx?Page=2

Acknowledgments

Daniel Sieberg

Eternal gratitude to my extraordinary parents, Geraldine Laundy and Douglas Sieberg, and to my inspirational sister, Jennifer Paronen, along with their respective and influential spouses, Robert Laundy, Mary Steel, and Lars Paronen. I'm blessed with a rich life surrounded by wise humans.

To my two brilliant, brave, and beautiful treasures—Kylie and Skye—every day you help me strive to leave a legacy worthy of being your father. And to my clever nephew Viggo: always explore the unknown.

To my divine life partner and confidant, Natalie Turvey—your unconditional love gives me wings without letting me fly too close to the sun. Always and forever.

Thank you to my former director at Google, friend and colleague, Scott Levitan, for opening the door for me to join GoodTrust at a time when I needed greater purpose and Ashlee Hunt at GoodTrust for your research skills. And to all of the Googlers/Xooglers and others around the world who believe in the power of technology for good.

Thank you to Dina Santorelli for your shared wisdom and editorial prowess. And to Mary Reynics at Penguin Random House for giving me my first platform as an author and for your undying encouragement. And to Ellen Scordato for welcoming us into the Stonesong publishing family.

Thank you to Joy Tan at Huawei USA for your faith, direction, and support.

For my friends, family, and former colleagues who have listened to my rambling tales of life, midlife crisis/opportunity and always helped me orient my compass to True North: Aaron Fyke, Rob Shaer, Rafael Jimenez, Dave Kubrak, Roger Patterson, Nancy Han, Rob Chestney, Geoff and Lindsay Loomer, Sean Carlson, Brendan Collins, Nick Whitaker, Erica Anderson, Shelagh McIntyre, Bill MacKenzie, Alex Walker, Anthony Laudato, Porter Anderson, Darwin Conner, Anita Kauser, Lisa Sauer, Kate Tobin, Marsha Walton, Miles O'Brien, Peter Dykstra, Steve and Mary Grove, Jesse Friedman, Olivia Ma, LaToya Drake, Ryan Bruno, Arun Venkataraman, Matt Cooke, Simon Rogers, Jen Bloch, Jessica Powell, Madhav Chinnappa, Kevin Allocca, Jeff Koyen, Nick Thompson, Joanna Stern, Will Schwalbe, Carmen Rita Wong, Nathan Long, Vadim Levitin, Susan Zirinsky, Karen Raffensperger, Jack Renaud, Lance Ulanoff, Marc Saltzman, Andrea Koppel-Pollack, Yarrow Kraner, Steve Cummins, Stephen Kirkpatrick, Stephen Ward, Mary Lynn Young, Graham Rockingham, Shelley Fralic, Patricia Graham, Jeff Jarvis, Beena Ammanath, Ana Jakimovska, Randall Jensen, Harminder S. Dhillon, and so many more. May your digital and real legacies live on. And never lose your infectious curiosity.

For Donna Logan who took a chance on me at the University of British Columbia's journalism school twenty-two years ago when e-mail was still a novelty.

For my dear uncle George Reginald MacKenzie (1936–2020)—gentle of mind and generous of spirit.

Thank you, dear reader, for the kindness of sharing your time with us.

To Rikard: let's do this. I will always be grateful for the opportunity to unite in service of something bigger than us. Thank you for your friendship, your counsel, and your good trust.

Rikard Steiber

Forever grateful for the support of my loving wife, Annika Steiber, and the inspiration from my two beautiful, intelligent, and curious daughters, Alexandra and Athena. To Paloma Picasso, our always happy and loyal dog. Without them, nothing would be possible or very meaningful.

To my wonderful parents, Maud Steiber and stepfather PG Peterson. To my brother, Viktor Peterson, who has always been a loyal friend. To my two sisters, Hilda and Anna, who always give me new perspectives and keep me grounded. To my father, John McLaren, who passed away this year and through his passing sparked the idea to write this book and launch the company GoodTrust.

A special thank you to Daniel Sieberg for making this book happen! Thank you to Dina Santorelli who edited the book into shape. And to Dave Brinda and David Romero for a great cover. Thank you Ellen Scordato at Stonesong publishing who helped us bring this book to market.

Thank you to the GoodTrust team and advisors who joined me on an exciting start-up journey: Markus Thorstvedt, Christian Lagerling, Olivia Gorajewski, Bjorn Laurin, Scott Levitan, Daniel Sieberg, Bjorn Book-Larsson, Ashlee Hunt, Axel Helmertz, Bastian Barsoe, Viktor Peterson, Markus Krieger, Terry Quan, Dave Brinda, Mario Verduzco, Robert Kullgren, Dennis Huang, Gopi Kallayil, Tony Fagan, Mike Beck, Shoshana Ungerleider, Sierra Campbell, Betsy Ehrenberg, Davis Slonim and Lee Poskanzer.

To my GoodTrust investors who provided me with great ideas, advice and resources to build a new business in the digital management space: Ben Ling, Nikesh Arora, Margo and Pete Georgiadis, Christian Wiklund, Bobby Lo, Scott Levitan, Warren Levitan, Arjan Djik, Tony Fagan, Gopi Kallayil, Jori Pearsall, Elisabeth Fullerton, Markus Thorstvedt, Björn Book-Larsson, Christian Lagerling, and Olivia Gorajewski.

To friends in the press that asked my important questions and highlighted the need to a better digital legacy solution in their media: Arielle Pardes (Wired), Dean Takahashi (VentureBeat), Charlie Fink (Forbes), Mark Saltzman (Tech It Out), Ed Butler (BBC), Johanna Ekström (Breakit), Martin Coulter (Business Insider).

To my friends who listened and gave me feedback when I pitched them GoodTrust and this book idea: Dean Gardner, Nicolai Wadström, Ben Levy, Nils Welin, Christian Wiklund, Jörgen Ericsson, Jonas Flodh, Erik Ingelsson, Maria Ingelsson, Nic Person, Denise Person, Anders Böös, Karen Wickre, Jessica Powell, Margo Georgiadis, Allan Thygesen, and Michael Rucker.

Dear reader, thank you for making this book effort worthwhile!

About the Authors

Daniel Sieberg is VP of Technology and Innovation Thought Leadership for Huawei USA and cofounder and content/PR lead for GoodTrust. He previously spent nearly three years in entrepreneurship developing companies across AI, AR, blockchain, and news aggregation. He also worked with nonprofits like First Street Foundation, which is building the first U.S. database of flood risk and served as a volunteer media relations specialist for the New York chapter of the American Red Cross. Prior to that, Sieberg was a group product marketing manager and official spokesperson at Google (2011–2017) and helped build two teams in service of journalism and technology.

During his time at Google he traveled to more than twenty-five Google offices, launched multiple large-scale projects, and is a graduate of the LEAD leadership program. As a correspondent and reporter for nearly fifteen years, Sieberg covered science and technology for ABC News, CBS News, and CNN, and began his career as a daily reporter with the *Vancouver Sun*. He has been nominated for five national News & Documentary Emmy Awards, and he has appeared as a featured guest with the likes of PBS News, MSNBC, BBC News, and NBC's Today Show; he also hosted dozens of episodes of CNN's weekly sci-tech show *NEXT@CNN* and forty episodes of *G Word* for Discovery Channel's Planet Green.

Sieberg is the author of the book *The Digital Diet: The Four-Step Plan to Break Your Tech Addiction and Regain Balance in Your Life*

(Crown, 2011), and he has written numerous articles on technology for publications including the *Washington Post*, *Details*, CNN.com, and others.

Sieberg holds a BFA (writing, film) from the University of Victoria (UVic) and an MJ (journalism, technology) from the University of British Columbia (UBC), and sits on the Board of Trustees at Saybrook University. A dual citizen of Canada and the United States, Sieberg has traveled to more than sixty-five countries and received the Chief Scout Award of Canada. He resides in Brooklyn with his two daughters and humbly requests that his headstone or digital memorialization feature the spinning pinwheel.

Rikard Steiber is a visionary Silicon Valley technology leader who recently launched his latest venture, GoodTrust, with the mission to protect everyone's digital legacy.

Steiber is an active senior adviser to Ericsson (telecoms) and Einride (autonomous electric vehicles), angel investor (in AI, VR, digital), and founder of several technology for good initiatives such as WIT Global—Europe's largest Women in Tech event and community. Steiber has a passion for space exploration and is a Future Astronaut with Virgin Galactic.

Steiber was a pioneer in the Virtual Reality industry where he, as president of Viveport and SVP of Virtual Reality at HTC, built Viveport, which is the world's first and only "Netflix of VR" subscription service live in more than sixty markets for all VR devices.

Before joining HTC, Steiber was the CEO for MTGx and chief digital officer at Modern Times Group, where he successfully built up several international businesses including Viaplay, a leading European Netflix competitor, several YouTube networks (with some 65,000 YouTube influencers) and a world-leading eSports player via acquisitions (including ESL and Dreamhack).

Steiber spent over six years at Google as the global marketing director of Ads, Mobile and Social Advertising, launching Google's most profitable and successful products. Prior to this, he ran marketing for all of Google's products in Europe, the Middle East, and Africa, launching groundbreaking consumer and business products such as Search, Maps, Android, YouTube, Gmail, Adwords, Adsense, and DoubleClick.

He cofounded Scandinavia Online (AOL of Nordics) and built Xlent Strategy Consulting into a leading digital/CRM strategy consultancy firm, helping leading media and telecom clients with their digital transformation.

Rikard has an MSc from SDA Bocconi in Italy and a BSc from Chalmers University of Technology in Sweden.

Made in the USA
Monee, IL
12 April 2021

65483603R00069